The Future
Just Happened

The Future Just Happened

Michael Lewis

Hodder & Stoughton

First published in Great Britain in 2001 by Hodder and Stoughton
A division of Hodder Headline

2 4 6 8 10 9 7 5 3 1

ISBN 0 340 79500 X

Typeset in New Baskerville
Printed and bound in Great Britain by
Mackays of Chatham plc, Chatham, Kent

Hodder and Stoughton
A division of Hodder Headline
338 Euston Road
London NW1 3BH

For Quinn

When we look at revolutions, we find that the outward acts
against the old order are invariably preceded by
the disintegration of inward allegiances
and loyalties. The images of kings topple
before their thrones do.

—PETER L. BERGER,
Invitation to Sociology

CONTENTS

THE INVISIBLE REVOLUTION

W hen Internet stocks began their freefall in February 2000, the Internet was finally put in its proper place. Everyone at once forgot that the omniscient chairman of the U.S. Federal Reserve, Alan Greenspan, had said that the Internet was changing the economy in ways that even he didn't fully understand. Or that Jack Welch, the CEO of General Electric, the closest thing to a multinational corporate monument, had said that the Internet was the most important force to strike the global economy since the Industrial Revolution. Or that the world's largest software company, Microsoft, was still saying that it needed to reinvent itself as an Internet business. Or that most large companies, even those that had nothing to do with high technology, were still frantically trying to figure out how to respond to the Internet. All of a sudden the Internet was just another technology, less important than the steam engine, the cotton gin, the telegraph, or air-conditioning. It was nothing more than a fast delivery service for information—that was what serious people who had either lost a lot of money in the late stages of the Internet boom or, more likely, failed to make money, liked to say now: "All the Internet does is speed up information—that's all."

Marshall McLuhan famously said that new technologies tend to become less visible as they become more familiar. The Internet was now proving his point. It was as if some crusty old baron who had been blasted out of his castle and was finally having a look at his first cannon had said, "All it does is speed up balls—that's all." The profit-making potential of the Internet had been overrated, and so the social effects of the Internet were presumed to be overrated. But they weren't. It is wildly disruptive to speed up information, and speeding up information was not the only thing the Internet had done. The Internet had made it possible for people to thwart all sorts of rules and conventions. It wasn't just the commercial order that was in flux. Many forms of authority were secured by locks waiting to be picked.

At any rate, I found the material too rich to ignore. Working on a book about a man at the epicenter of the Internet Boom, I had stumbled upon a lot of disturbing Internet-inspired events: children who had used the new tool to become financial experts; parents who had used it to cede the responsibility for knowing about the world; big businessmen who had used it to transform themselves into enemies of the mass market; antisocial technologists who had been encouraged by it to reinvent themselves as social theorists. The technology of the Internet was far less interesting than the effects people were allowing it to have on their lives, and what these, in turn, said about those lives. What was happening on the Internet buttressed a school of thought in sociology known as role theory. The role theorists argue that we have no "self" as such. Our selves are merely the masks we wear in response to the social situations in which we find ourselves. The Internet had offered up a new set of social situations, to which people had responded by grabbing for a new set of masks. What was true of people was seemingly also true of ideas, as capitalism itself suddenly appeared willing to take new risks with its identity. I had already seen enough of the identity free-for-all to know that it was deeply unsettling. I could only assume that there were many more disturbing Internet-inspired events that I didn't know about.

And that was the problem. By the spring of 2000 hundreds of millions of people were on line. They had created billions and billions of web pages. The best Internet search engines didn't reach more than about a fifth of these. The best Internet search engine did not even know how many web pages existed. There was no way I was going to be able to investigate this vast new social world by myself. I needed help looking around. Into the picture strolled the British Broadcasting Corporation. It turned out that the BBC had already set aside a pile of money to make a television series about the social consequences of the Internet. This enabled them to hire a team of pro web surfers to help me scour the Internet for telling examples of human perversion. When they found something worth seeing, they would tell me about it. Then we would all go out together and knock on the front doors of the people behind the events and see what they were like in the flesh.

The British people had generously offered to fund an investigation. But of what? Certainly not the definitive catalogue of "the social consequences of the Internet." It's far too early for that. In the long run the Internet will become invisible and ubiquitous, and no one will spend a minute thinking about its social effects any more than they now think about the social effects of electricity. What I was after was more like the Internet consequences of society. People take on the new tools they are ready for, and only make use of what they need, how they need it. If they were using the Internet to experiment with their identities, it was probably because they found their old identities were inadequate. If the Internet was giving the world a shove in a certain direction, it was probably because the world already felt inclined to move in that direction. When I realized this I stopped worrying over the social consequences of the Internet and began simply to watch what was actually happening on the Internet. Inadvertently, it was telling us what we wanted to become.

It had an odd way of doing this, however. After a few months on the road I realized that I was spending a lot of my time chasing after children. This was new. I was accustomed to being younger than my subjects. All of a sudden I was the weird old guy who hangs around outside the school gate and waits for the bell to ring. I was uneasy in this new role. It seemed a truly perverse way to waste a lot of time. It wasn't until I traveled to Finland that I realized that there was a deeply serious commercial precedent for an unseemly interest in children. Oddly enough, a Finnish company, Nokia, had come to dominate the mobile phone business to the point where pretty much everyone now agreed that the

Finns would be the first to connect mobile phones to the Internet in a way that the rest of us would find necessary. If one day we all wound up walking around physically attached to the Internet—and it seemed likely that at least some of us would—Nokia would be the immediate reason for it. Overnight the Finns had gone from being celebrated mainly for their tendency to drink too much and then kill themselves to being heralded as the geniuses who built the most advanced communications industry on the planet. They had done this in spite of being personally uncommunicative, the only people I have ever met who, as they became drunk, grew even more silent.

The Finns were successful because they were especially good at guessing what others would want from their mobile phones. One big reason for this—or so the people at Nokia believed—was that they spent a lot of time studying children. The kids came to each new technology fresh, without preconceptions, and they picked it up more quickly. They dreamed up uses for their phones that, for reasons no one fully understood, never occurred to grown-ups. The instant text message, for instance. The instant message was fast becoming a staple of European corporate communication. To create an instant message, you punched it by hand into your telephone, using the keypad as a typewriter. On the face of it this is not an obvious use of a telephone keypad. The difference between the number of letters in the alphabet and the number of keys on the pad meant you wound up having to type a kind of Morse code. The technique had been invented by Finnish schoolboys who were nervous about asking girls out on dates to their face, and Finnish schoolgirls who wanted to tell each other what had happened on those dates, as soon as it happened. They'd proved that if the need to communicate indirectly is sufficiently urgent, Words can be typed into a telephone keypad with amazing speed. Five and a half million Finns had sent each other more than a billion instant messages in the year 2000. The technique had spread from Finnish children to businessmen because the kids had taught their parents how to use their phones. Nokia employed anthropologists to tell them this.

Finland had become the first nation on earth to acknowledge formally the child-centric model of economic development: if you wanted a fast-growing economy, you needed to promote rapid technical change, and if you intended to promote rapid technical change, you needed to cede to children a strange measure of authority. The average twelve-year-old Finn now owned a mobile phone, and it was widely assumed inside Nokia that one day every seven-year-old Finn would own a mobile phone. The twelve-year-olds disapproved of this—they'd crinkle their foreheads and say that, really, seven was too young. They didn't understand that their futures depended on seven-year-olds having phones. If the twelve-year-olds were able to transform business communication, who knew what the seven-year-olds might achieve?

I don't want to dwell here on why children—and, more generally, childishness—plays such an important part in the story that follows. I'll dwell on that enough later. But it does seem to me that when capitalism encourages ever more rapid change, children enjoy one big advantage over adults: they haven't decided who they are. They haven't sunk a lot of psychological capital into a particular self. When a technology comes along that rewards people who are willing to chuck overboard their old selves for new ones—and it isn't just the Internet that does this; biotechnology offers many promising self-altering possibilities—the people who aren't much invested in their old selves have an edge. The things that get tossed overboard with a twelve-year-old self don't seem like much to give up at the time. Some part of the following story required me to remember this cruel fact. When you find that you are now the weird old guy hanging around outside the school gates, it becomes a necessary act of will to recall what it feels like to be a child, in the process of ordering up a self from the menu. It is necessary to recall, in particular, how ruthless the process can be.

I had spent my childhood in New Orleans. I would like now to consider this otherwise uninteresting fact as it is bound up with my interest in what follows. By the mercenary standards of the modern world, New Orleans is a failed place. In my lifetime it has ceased to be the capital of trade and commerce in the American South and become a museum city, like Venice. The new capital of the American South is Atlanta, which has made the shrewd but spiritually vacuous decision not to stand on ceremony or tradition but rather to go whoring after progress. Atlanta has transformed itself. It is no longer even a city; it's an airport, a blur of movement unrelated to anything but the pursuit of money. It is also, not uncoincidentally, one of America's Internet business centers.

Not so New Orleans. Decades of economic failure are in many ways unappealing, but in one way they are an advantage. Where there is no economic development there is no big change. There is just a slow, inexorable crumbling. For that reason New Orleans has always been an excellent place to observe progress. (The same might be said for any number of European cities: Manchester, England; Paris, France.) To know progress you need to know what it has rolled over or left behind, and when progress is moving as fast as it now is, recalling its victims is difficult. New Orleans keeps its anachronisms alive long enough for them to throw the outside world into sharp relief. For instance, until the mid-1990s you could find actual gentlemen lawyers in New Orleans, who thought of themselves mainly as members of an honorable and dignified profession. One of these dinosaurs was my father.

Right up until it collapsed, the old family law firm that my father managed clung to its charming habits. The gentlemen lawyers wrote notes to each other arguing over the correct pronunciation of certain phrases in ancient Greek. They collected strange artifacts from dead cultures. They treated education as a branch of religion. They wore bow ties. They were terrifyingly at ease with themselves but did not know the meaning of casual Friday. Their lives had been premised on a frankly elitist idea: an attorney was above the fray. He possessed special knowledge. He observed a strict code of conduct without ever having to say what it was. He viewed all entreaties to change with suspicion. (The lawyer in the office next door to my father not only shunned e-mail when it arrived; he still used a telephone from 1919 that had belonged to his father.) The most important thing in the world to him was his stature in the community, and yet so far as anyone else could determine he never devoted an ounce of his mental energy to worrying about it. Status wasn't a cause; it was an effect of the way he led his life.

The first hint I had that this was no longer a tenable pose—and would not be a tenable pose for me—came from a man I'd never met called Morris Bart. I was some kind of teenager at the time. My father and I were driving along the Interstate highway that ran through town when we came upon a giant billboard. It said something like ARE YOU A VICTIM? HAVE YOU BEEN INJURED? NO ONE REPRESENTS YOUR INTERESTS? CALL MORRIS BART: ATTORNEY-AT-LAW. And there was a big picture of Morris Bart. He had the easy smile of a used car dealer.

"Do you do the same thing as Morris Bart?"

"Not exactly."

"But his billboard says he's a lawyer."

"We have a different kind of law firm."

"How?"

"We don't have billboards."

"Why not?"

"It's just not something a lawyer does."

That was true. It was true right up to the moment Morris Bart stuck up his picture beside the Interstate highway. My father and his colleagues remained unmoved, but the law was succumbing to a general force, the twin American instincts to democratize and to commercialize. (Often they amount to the same thing.) These are the two forces that power the Internet, and in turn are powered by it. Martin Sorrell, the British chairman of the global advertising firm WPP, says there is no such thing as globalization. There is only Americanization. I know a few French chefs and German car manufacturers and even British advertising executives who would dispute that statement. But the man has a point. And I know what it feels like to be on the wrong end of the trend. New Orleans knew how the world outside of America felt about America because New Orleans felt that way too.

Morris Bart was a tiny widget inside the same magnificent American instrument of destruction that the Internet has so eloquently upgraded. A few years after he put up his billboard my father's firm began to receive calls from "consultants" who wanted to help them learn how to steal clients and lawyers from other firms—a notion that would have been unthinkable a few years earlier, and remained unthinkable to some. A few years after that the clients insisted that lawyers bill by the hour—and then questioned the bills! The old game was over. The minute the market intruded too explicitly, the old prestige began to seep out of the law. For the gentlemen lawyers it ended about as well as it could. But still it ended. And for people whose identity was wrapped up in the idea, the end gave their story the shape of tragedy.

The gentlemen lawyers responded to the assault on their world in character, by refusing to give an inch. Their children responded differently. A child still has time to save himself. To a child, being on the wrong end of the trend is not a sign that it's time to dig in and defend the old position; it's a signal to cut and run. Progress depends on these small acts of treason.

I recall the feeling when it first dawned on me that the ground beneath my teenage feet was moving. I did not enjoy the premonition of doom in my father's world. But what troubled me even more was that some part of me wanted my father to have his own billboard beside the highway—which of course he would never do. My response was to leave home and invent another self for myself. Had the Internet been available, I might have simply gone on line.

O N E

THE FINANCIAL REVOLT

One day some social historian will look back with wonder on the havoc wreaked by the Internet. The events on Wall Street alone will offer him material for an entire chapter. Buried in the footnotes to that chapter will be the wry little anecdote about the first child to manipulate the stock market. The two great forces that propel mankind forward—anecdote and accident—are not historically respectable, so that's all this story will get: a footnote. It deserves much more than that.

On September 21, 2000, the U.S. Securities and Exchange Commission settled its case against Jonathan Lebed. The SEC's press release explained that fifteen-year-old Jonathan—the first minor ever charged with stock market fraud—had used the Internet to promote stocks from his bedroom in Cedar Grove, New Jersey. Armed only with accounts at AOL and E-Trade, the kid had bought stock, then, using "multiple fictitious names," posted hundreds of messages on Yahoo Finance message boards recommending that stock to others. He'd done this eleven times between September 1999 and February 2000, the SEC said, each time triggering chaos in the stock market. In advance of the chaos he'd left sell orders in the marketplace, in case his shares rose in price. Each time they did. The average daily trading volume of the small companies he dealt in was about 60,000 shares; on the days he posted his messages volume soared to more than a million shares. More to the point, he'd made money. Between September 1999 and February 2000 his smallest one-day gain was $12,000. His biggest was $74,000. Now the kid had agreed to hand over his illicit gains, plus interest, which came to $285,000.

When I first read the newspaper reports I didn't understand them. It wasn't just that I didn't understand what the kid had done wrong. I didn't even understand what he had done. And if the first news stories about Jonathan Lebed raised questions—What did it mean to use a "fictitious name" on the Internet, where every name is fictitious? Who were these people who traded stocks naïvely based on what they read on the Internet?—they were trivial next to the questions raised a few days later, when a reporter asked Jonathan Lebed's lawyer if the SEC had taken all of the profits. They hadn't. There had been many more than the eleven trades described in the SEC's press release, the lawyer said. The kid's take from six months of trading had been nearly $800,000. Initially the SEC had demanded he give it all up, then backed off when the kid put up a fight. As a result, Jonathan Lebed was still sitting on half a million dollars, which he'd made in less than six months of trading.

At length, I phoned the Philadelphia office of the SEC, where I reached one of the investigators who had brought Jonathan Lebed to book. I was maybe the fiftieth journalist he'd spoken with that day, and apparently a lot of the others had had trouble grasping the finer points of securities law. At any rate, by the time I asked him to explain to me what, exactly, was wrong with broadcasting one's private opinion of a stock on the Internet, he was in no mood.

"Tell me about the kid," I said.

"He's a little jerk."

"How so?"

"He is exactly what you or I hope our kids never turn out to be."

"Have you met him?"

"No. I don't need to."

O n one of those gray winter mornings that make New Jersey seem even more unfairly like New Jersey, I landed at Newark Airport, where I was met by a lawyer named Kevin Marino. Marino was on his way out the door of his Newark office on April 5, 2000, when Jonathan Lebed walked in with his father. Marino had agreed to see the Lebeds as a favor to a mutual friend, but he didn't have much interest in their bizarre case, or in any case. He'd been working too hard and hadn't seen enough of his wife and two sons and so had decided to take a long vacation out West. But before he did that, to placate his old friend, he spent an hour with Jonathan and Greg Lebed of Cedar Grove, New Jersey.

In that hour a couple of things became clear. The first was that Greg Lebed, though outraged that the government was demanding that his son hand over $800,000, still didn't fully understand what his son had done to accumulate it. The other was that Jonathan Lebed, who didn't appear to much care what became of his money, knew more than he was saying. "Greg didn't understand what Jonathan was up to and Jonathan wasn't saying what Jonathan was up to," says Marino. "I had a sense he wasn't buying what anyone was telling him, but he wouldn't say anything. So I ask him: 'What's it like to be dragged in and cross-examined by the government? What were *they* like?' And he made this face like he didn't want to say that they didn't have a clue, but that's what he's thinking. So I asked him: 'What did you think of their questions?' And he thinks about that for a minute and then says, deadpan, 'Some of them were good.'"

Fifteen years of defending people accused of white-collar crimes had left Marino essentially incapable of being shocked by human behavior. But when he comes to speak of the case at hand—how Jonathan Lebed came to be a poster boy for Internet stock fraud—his voice admits wonder. "I had a very clear sense right away that there was something dramatic and unusual about him. For a kid his age he had *huge* nerve."

Cedar Grove, New Jersey, was one of those Essex County suburbs defined by the fact that it was not Newark. The real estate prices appeared to rise with the hills. The houses at the bottom of each hill were barely middle class; the houses at the top might fairly be described as opulent; but in some strange way they were all the *same* house. Even million-dollar homes built on streets with names like Tiffany Court were less upper-class mansions than some middle-class person's idea of upper-class mansions. Indistinguishable from the homes on either side of them—same manicured lawn, same grandiose entryway, same more-crystal-than-crystal chandeliers—they were, in essence, giant tract houses. In Cedar Grove rich just means having more of exactly what you had when you weren't rich.

About a third of the way up the first hill, Marino stopped and turned left. For most of the ride down from Newark he'd been describing his seriocomic negotiations with the SEC, which had ended with a phone call from an SEC lawyer who said he was "surprised and disappointed" that the media had discovered the $550,000 Jonathan Lebed had kept. Marino had just wanted to keep as much money as possible for his client; the SEC was more concerned with maintaining the illusion that crime didn't pay. When news broke that the kid was still sitting on more than half a million dollars, the illusion broke with it. Now Marino slowed the car and said, "You are about to see the Sticking Point." Before I had time to ask what he meant we had turned into a driveway and I was able to see what he meant: parked in front of a distinctly modest home, hood ornament out, was a bright green Mercedes SUV. Just before the SEC called him, Jonathan Lebed had taken $41,000 of his profits and bought a car. He was still too young to drive the car himself but he enjoyed being driven around in it. "That was the sticking point in my negotiations with our government," says Marino. "The SEC couldn't get past the Mercedes. It still ticks them off."

The first person to the door that day was Greg Lebed. Jonathan's father cut a figure. Black hair sprouted in many directions from the top of his head and joined together deep in the middle of his back. The "Ape Drape." The curl of his lip seemed designed to shout abuse from a bleacher seat. Dress him up as an Eskimo and drop him into the Arctic Circle and he'd remain, unmistakably, Essex County. Anyway, that was the first impression he made, of a man perfectly untamed, a creature at home only in the New Jersey wild. He'd furthered this impression in his brief encounters with the mass media. As he'd tossed fifty or so journalists off his front lawn, he'd declared, "I'm proud of my son." That hit the newspapers and elicited no end of moral indignation from the SEC. Poked and prodded about the comment by an interviewer from *60 Minutes,* he'd gone on to say, "It's not like he was out stealin' the hubcaps off cars or peddlin' drugs to the neighbors."

But when he wasn't busy being outraged or indignant, Greg was unfailingly polite, almost meek. He apologized to me for taking so long to answer the door. He was down in his basement, he said, fiddling with his trains. That's mainly what he did when he was home: played with his model train set in the basement. You'd be upstairs fixing a cup of tea, enjoying the tranquillity of Cedar Grove, when Greg would take his vintage B&O Railroad engine out for a spin around the basement tracks and all of a sudden there was a shriek and a whistle that made the house feel like it was built on top of Penn Station. But I didn't learn that until later. On that winter afternoon Greg led us to the family dining room and, without the slightest help from me, worked himself into a fury. "Youse want to see something!" he hollered. "Let me just shows youse something!" He pulled out a photocopy of the front page of the New York *Daily News.* One side displayed a snapshot of Bill and Hillary Clinton beside the headline CLINTONS CLEARED: INSUFFICIENT EVIDENCE ON WHITEWATER SAYS REPORT; the other side had a picture of Jonathan Lebed beside the headline TEEN STOCK WHIZ NAILED. Over it all, in Greg's furious hand, was scrawled: "U.S. Justice at Work."

"Look at that!" Greg shouted. "This is what goes on in this country!"

Then, as suddenly as he'd erupted, he went dormant. "Don't bother with me," he said, "I get upset." He offered seats at his dining room table. From that moment his anger ceased to be frightening, and became something else.

Connie Lebed now entered. Jonathan's mother had a look on her face that as much as said, "I assume Greg has already started yelling about something. Don't mind him, I certainly don't." When her husband yelled, she was as good as deaf.

"It was that goddamn computer what was the problem," said Greg, testily.

"My problem with the SEC," said Connie, ignoring her husband, "was that they never called. One day we get this package from Federal Express with the whaddyacallit, the subpoenas inside. . . . If only they had called me first." She would say this six times before the end of the day, with one of those marvelous northern New Jersey harmonica-like wails that conveys a sense of grievance maybe better than any noise on the planet. *If only theyda caaaawwwwlled me.*

"The wife brought that goddamn computer into this house in the first place," Greg said, hurling a thumb at Connie. "Ever since that computer came into the house this family was ruined."

Connie absorbed the full frontal attack with an uncomprehending blink, and then, as if her husband had never spoken, said to me, "My husband has a lot of anger. He gets worked up easily. He's already had one heart attack." She neither expected nor received the faintest reply from him. They obeyed the conventions of the stage: whenever one of them stepped forward into the spotlight to narrate, the other receded, and froze like a statue. The U.S. government had made a point of asking Greg how he got on with what he invariably called "the wife," so I didn't need to.

> SEC: *How would you describe your relationship with your wife, just generally?*
>
> GREG: *So, so. There have been a lot of problems lately, such as this.*
>
> SEC: *I mean do you communicate and talk?*
>
> GREG: *Oh, yes, yes.*
>
> SEC: *You discuss things?*
>
> GREG: *Yes.*
>
> SEC: *But this whole thing that we are investigating has created some tensions?*
>
> GREG: *Yes.*
>
> SEC: *Anything abnormal?*
>
> GREG: *Just a lousy period. In other words, it ain't a happy thing, you know.*
>
> SEC: *Well, who was angry about it, you or her or both?*
>
> GREG: *Well, it upset me.*
>
> SEC: *Right.*
>
> GREG: *And, of course, it upset her.*
>
> SEC: *Did you have any sit-downs with Jonathan, the three of you?*
>
> GREG: *Yes.*
>
> SEC: *Prior to [your] contact with the SEC, was your relationship with your wife better?*
>
> GREG: *Yes.*
>
> SEC: *And you communicated more?*
>
> GREG: *Yes.*

Ten minutes into that first conversation at the Lebed dining room table Jonathan slouched in. But even that verb does not capture the mixture of sullenness and truculence with which he entered the room. He was long and thin and dressed in the prison costume of the American suburban teenager—pants too big, sneakers gaping, a pirate hoop dangling from one ear. He looked away when he shook my hand, and said "Nice to meet you" in a way that made it clear he couldn't be less pleased, then sat down and said nothing while his parents returned to their split-screen narration. At a glance, it was impossible to link Jonathan in the flesh to Jonathan on the Web. I had a fat file of the collected Internet postings of Jonathan Lebed and they were marvelously bombastic. For instance, just two days before the FedEx package arrived on the Lebeds' front doorstep bearing the SEC's subpoenas, Jonathan had logged onto the Internet and posted two hundred separate times the following plug for a company called Firetector (ticker symbol FTEC):

SUBJECT: **The Most Undervalued Stock Ever**

DATE: 2/03/00 3:43pm Pacific Standard Time

FROM: LebedTG1

FTEC is starting to break out! Next week, this thing will EXPLODE . . .

Currently FTEC is trading for just $2½. I am expecting to see FTEC at $20 VERY SOON . . .

Let me explain why . . .

Revenues for the year should very conservatively be around $20 million. The average company in the industry trades with a price/sales ratio of 3.45.

With 1.57 million shares outstanding, this will value FTEC at . . . $44.

It is very possible that FTEC will see $44, but since I would like to remain very conservative . . . my short term price target on FTEC is still $20!

The FTEC offices are extremely busy . . . I am hearing that a number of HUGE deals are being worked on. Once we get some news from FTEC and the word gets out about the company . . . it will take-off to MUCH HIGHER LEVELS!

I see little risk when purchasing FTEC at these DIRT-CHEAP PRICES. FTEC is making TREMEN-DOUS PROFITS and is trading UNDER BOOK VALUE!!!

This is the #1 INDUSTRY you can POSSIBLY be in RIGHT NOW.

There are thousands of schools nationwide who need FTEC to install security systems . . . You can't find a better positioned company than FTEC!

These prices are GROUND-FLOOR! My prediction is that this will be the #1 performing stock on the NASDAQ in 2000. I am loading up with all of the shares of FTEC I possibly can before it makes a run to $20.

Be sure to take the time to do your research on FTEC! You will probably never come across an opportunity this HUGE ever again in your entire life.

The author of that and dozens more like it now sat dully at the end of the family's dining room table and watched his parents take potshots at each other and their government. There was not an exclamation point in him. Clearly just trying to get his client interested in the conversation for the benefit of an outside observer, his lawyer, who had been watching in silence what was to him a familiar scene, recalled the day the SEC handed over Jonathan's name and address to the press. He called to tell the Lebeds that they should stay inside and not have anything to do with journalists, and certainly not answer any questions. On cue, journalists from just about every place on the planet overran the Lebeds' front yard. News trucks. Fat guys with huge cameras. Screeching women in sneakers. Connie called Greg and told him she wanted to move away from Cedar Grove, to a place where no one knew them. Their eleven-year-old daughter, Dana, burst into tears and ran out the back door with the promise that she was going to live the rest of her life at a friend's house. Greg hollered expletives into the phone and hopped on a high-speed train to Triple Bypass. Jonathan was the only one who remained calm. While Connie spoke to the family lawyer—whose card was now glued to the wall over the phone—Jonathan walked out the front door and presented himself to the mob.

"Jonathan," Connie wailed. "Kevin said not to go out there. Wheyougoin'?"

"What's going on there?" Marino shouted.

"Jonathan's outside holding a press conference!!!"

That's just what he had done. He had strolled out the front door and told the reporters that he would talk to them on condition that they mention the name of his new web site in their newspapers.

Connie now said, "I'm inside yellin', 'JON-A-THAN!

Come in! Come in! Kevin says come in!' And he's not comin' in. So I take the phone out to the front and make him talk to Kevin. Kevin yelled at him to get in the house."

"It was so fun that day," Jonathan said, then ducked back inside his shell.

It took another twenty minutes for their attorney to drag the family together into the general discussion of Jonathan Lebed's life and works, and another few weeks of me pestering them to get a general answer to the question: How on earth did Jonathan Lebed happen? The story finally came out in full over dinner one night. In its short version, it ran roughly as follows:

Not long after his eleventh birthday, Jonathan opened an account with America Online. He went onto the Internet, at least at first, to meet other pro wrestling fans. He built a web site dedicated to the greater glory of "Stone Cold" Steve Austin. But at about the same time, by watching his father, he became interested in the stock market. In his thirty-three years working for Amtrak, Greg Lebed had worked his way up from a simple laborer to middle manager. Along the way he'd accumulated maybe twelve thousand dollars of blue-chip stocks. Like half of America, he'd come to watch the market's daily upward leaps and jerks with keen interest. Jonathan saved him the trouble. When he'd come home from school, he'd turn on CNBC and watch the stock market ticker stream across the bottom of the screen, searching it for the symbols inside his father's portfolio.

"Jonathan would sit there for hours staring at them," said Connie, as if Jonathan were miles away.

"I just liked to watch the numbers go across the screen," Jonathan said.

"Why?"

"I don't know," he said. "I just wondered, like, what they meant."

At first the numbers meant a chance to talk to his father. He'd call whenever he saw one of his father's stocks cross the bottom of the television screen. This went on for about six months before Jonathan declared his own interest in owning stocks. On September 29, 1996, Jonathan's twelfth birthday, a savings bond his parents had given him at birth came due. He took the $8,000 and persuaded his father to invest it on his behalf in the stock market. The first stock he bought was America Online, at $25 a share—in spite of a lot of adverse commentary about the company on CNBC. "He said that it was a stupid company and that it would go to 2 cents," Jonathan said, pointing at his father, who obeyed what I came to understand as the family rule, and sat frozen at the back of some mental stage. AOL rose five points in a couple of weeks, and Jonathan sold it. From this he learned (a) you could make money quickly in the stock market, (b) his dad didn't know what he was talking about, and (c) it paid him to exercise his own judgment on these matters. All three lessons were reinforced dramatically by what happened next.

What happened next was that CNBC—which Jonathan now rose at five each morning to watch—announced its stock-picking contest for children. Jonathan had wanted to join the contest on his own but was told that he needed to be on a team, so he went and asked two friends to join him. Several thousand student teams from schools across the country, each consisting of at least three children, set out to speculate their way to victory. Each afternoon CNBC announced the top five teams of the day. To get your name read out loud on television you obviously opted for highly volatile stocks that stood a chance of doing well in the short term. Jonathan's team—which dubbed itself "the Triple Threat"—had a portfolio that rose 51 percent the first day, which put them in second place. They remained in the top three for the next three months, until, in the last two weeks of the contest, they collapsed. "The problem was that in the last two weeks I listened to my teammates," said Jonathan. "I did it all until the end and then I let them talk me into buying Netscape and Western Union. So we finished fourth."

Even a fourth-place finish was good enough to fetch a camera crew from CNBC, which came and filmed the team in Cedar Grove. The Triple Threat was featured in the *Verona-Cedar Grove Times* and celebrated on television by the Cedar Grove Township Council. "From then everyone at work started asking me if Jonathan had any stock tips for them," said Greg. "They *still* ask me," said Connie. From that time also dated Jonathan's most distinctive attribute: his attaché case. In it he kept reams of stock market research; and he never went to school without it. Everywhere Jonathan went he found grown-ups asking for advice about the markets. A couple of them were professional stockbrokers. "He was taking a physics test," said Connie, "and the biology teacher knocked on the door and pulled him out of class to ask him if he had any stock ideas."

"No," said Jonathan, "that teacher already had his stocks. He had a list of five stocks that he thought had bottomed out, and he wanted to know what I thought." He said this in an indifferent tone, as if it happened every day, because it did. The effect on Cedar Grove public opinion of the SEC's call was to confirm the general suspicion that Jonathan knew something other people did not.

By the spring of 1998 Jonathan's ambitions were growing. He was now thirteen. He had glimpsed the essential truth of the market—that even people who called themselves professionals were often incapable of independent thought, and that most people, though obsessed with money, had little ability to make decisions about it—and realized he was special. He knew what he was doing, or thought he did. He'd learned how to find everything he wanted to know about a company on the Internet; what he couldn't find he ran down in the flesh. It became a routine in Connie Lebed's life to drive her son to various corporate headquarters to make sure they existed. Considering an investment in a bagel company, Jonathan made her drive him to taste the bagels on the other side of New Jersey. When he wanted to invest in a cigar company, he attempted to take the entire family seventy-five miles out of its way on a vacation to Florida so that he could inspect the company's newest retail outlet.

Jonathan entered the next CNBC competition, but soon became bored and abandoned it after a few weeks. "What's the point of doing it with fake money?" he said. Instead his mother opened an account with Ameritrade. "He done so well with the stock contest, I figured, Let's see what he can do." said Connie.

What he did was turn his $8,000 savings bond into $28,000, inside of a year. At around the same time, he created his own web site devoted to companies with small market capitalization—penny stocks. The web site was called Stock-Dogs.com. ("You know, like racing dogs.") Stock-Dogs.com plugged the stocks of companies Jonathan found interest-ing, or that people Jonathan met on the Internet found interesting. At its peak Stock-Dogs.com had maybe fifteen hundred visitors a day. Even so, the officers of what seemed to Jonathan to be serious companies wrote to him to sell him on their companies. Within a couple of months of becoming an amateur stock market analyst, he was in the middle of a web of people who spent every waking hour chatting about and trading stocks on the Internet.

The mere memory of this clearly upset Greg even more than usual. We were sitting around the dining room table, in the middle of dinner, when he was forced to retrieve it.

"He was just a little kid!" he exploded. "These people who got in touch with him could have been anybody."

"How do you know?" said Jonathan. "You've never even been on the Internet."

"I started to get scared shit," Greg said. "Suppose some hacker comes in and steals his money! Next day you type in and you got nuttin' left."

Jonathan snorted. "That can't happen." He turned to me. "Whenever he sees something on TV about the Internet he gets mad and disconnects my computer phone line."

"Oh yeah," said Connie, brightening as if realizing for the first time that she lived in the same house as the other two. "I used to hear the garage door opening at three in the morning. Then Jonathan's little feet running back up the stairs."

"Youse goddamn right I unplugged it," said Greg. He turned to me to explain. "The phone line was in the garage." It didn't seem to bother him in the slightest that every time after he'd unplugged it, his son waited until he'd gone to bed, then ran down and plugged it back in. It didn't seem to occur to him that he might be able to control his child's actions. The point was: at least he'd done *something*.

"He'd see something about credit card fraud and he'd start shouting," said Jonathan. "He believes it, like, happens very commonly, and it doesn't. Anyway, if someone steals your credit card on the Internet, it's not your responsibility. He doesn't even know that."

"I haven't ever even turned a computer on!" said Greg. "And I never will!"

"He just doesn't understand how a lot of this works," explained Jonathan, patiently. "And so he overreacts sometimes."

"I know, I know," said Greg, turning to me. "I'm supposed to know how it works. It's the future. But that's *his* future, not mine!"

Greg and Connie were both born in New Jersey, but from the moment the Internet struck, they might as well have just arrived from Taiwan. When the Internet landed on them it had redistributed the prestige and authority that goes with a general understanding of the ways of the world away from the grown-ups and to the child. The grown-ups now depended on the child to translate for them. Technology had turned them into a family of immigrants.

"Anyway," said Connie, drifting back in again. "That's when the SEC called us the first time."

The first time?

The first time. Jonathan had met the authorities charged with maintaining the appearance of propriety in the capital markets once before, in late October of 1998. He'd just turned fourteen. The SEC did, at least the first time, call

Connie. It was against the law for a minor to open a broker-age account, and so the Ameritrade account Jonathan used to trade stocks remained in Connie's name. This raised the obvious suspicions. "They didn't believe me when I told them that Jonathan was making his own decisions," says Con-nie. "They kept asking about me and Greg."

"Yeah," says Greg. "They thought I was, like, encouraging him to break the law."

Frightened and worried, Connie agreed to bring her son in to meet with the SEC in its Manhattan offices. When he heard *that* news, Greg, of course, hit the roof. "He'd already had one heart attack," Connie explained, and began to review the heart problems all over again, inspiring Greg to mutter something about how *he* wasn't the person who brought the computer into the house and so it wasn't *his* responsibility to deal with this little nuisance.

So Connie asked Harold Burk, her boss at the drug com-pany Hoffmann-La Roche, where she temped, to go with her and Jonathan. Together they made their way to a long con-ference table in a big room at 7 World Trade Center. On one side of the table, five lawyers and an examiner from the SEC; on the other, a fourteen-year-old boy, his mother, and a bewildered friend. This is how it began:

> SEC: *Does Jonathan's father know he's here today?*
> MRS. LEBED: *Yes.*
> SEC: *And he approves of having you here?*
> MRS. LEBED: *Right, he doesn't want to go.*
> SEC: *He's aware you're here.*
> MRS. LEBED: *With Harold.*

SEC: *And that Mr. Burk is here.*

MRS. LEBED: *He did not want to—this whole thing has upset my husband a lot. He had a heart attack about a year ago and he gets very, very upset about things. So he really did not want anything to do with it and I just felt like—Harold said he would help me.*

The SEC quickly seems to have figured out that they were racing into some strange mental cul-de-sac, and that the mother hadn't the first notion of what her son had been up to. They turned their attention to Jonathan, or, more specifically, his brokerage statements.

SEC: *Can you flip to page 2 of the March 28th through April 24, 1998 statements. You'll see that on April 16th there are a couple transactions in a stock called Keytel International Inc.*

JONATHAN: *Yes.*

SEC: *It appears, Jonathan, that you bought and sold Keytel, you bought it twice and you sold it twice on that day and the sales were at a loss. Can you explain your trading in this stock for us?*

JONATHAN: *I was trying to day trade it for a profit with the momentum that it was having that day.*

SEC: *What do you mean when you say "momentum"?*

JONATHAN: *It was moving up very, very fast that day. I think it doubled that day or something because they announced that they would be selling their music on the Internet and stuff. So I just tried to buy it and sell it throughout the day to make money with it.*

SEC: *Is there any possibility that you were trying to con-tribute to some of that momentum by day trading?*

JONATHAN: *No, the volume was like real high. My trading wouldn't contribute at all.*

SEC: *What do you mean it wouldn't contribute at all?*

JONATHAN: *My trading—it was just like 1,000 shares and the volume for the day was in the millions. It was trading a lot, so it was not like it would contribute.*

SEC: *Are there any instances in which you've engaged in day trading in order to contribute to the momentum of the stock?*

JONATHAN: *No.*

SSEC: *Where did you learn your technique for day trad-ing?*

JONATHAN: *Just on TV, internet.*

SEC: *What TV shows?*

JONATHAN: *CNBC mostly—basically CNBC is what I watch all the time.*

SEC: *Do you generally make money on day trading?*

JONATHAN: *I usually don't day trade, I just try to—since I was home these days and I was very bored, I wanted something to do, so I was just trading con-stantly. I don't think I was making money. . . .*

SEC: *Just looking at your April statement, it looks like the majority of your trading is day trading.*

JONATHAN: *I was home a lot that time.*

MRS. LEBED: *They were on Spring vacation that week.*

Having established and then promptly ignored the boy's chief motive for trading stocks—a desire to escape the tedium of existence—the authorities then sought to discover his approach to attracting attention on the Internet.

> SEC: *On the first page [referring to a hard copy of Jonathan's web site, Stock-Dogs.com] where it says "Our six to twelve month outlook, $8," what does that mean? The stock is selling less than three but you think it's going to go to eight.*
>
> JONATHAN: *That's our outlook for the price to go based on their earnings potential and a good value ratio. . .*
>
> SEC: *Are you aware that there are laws that regulate company projections?*
>
> JONATHAN: *No.*

At length the SEC people snuck up on the reason they had noticed Jonathan in the first place. They'd been hot on the trail of a grown person named Ira Monas, one of Jonathan Lebed's many Internet correspondents. Monas, who was eventually jailed, had been employed in "investor relations" by a number of small companies. In that role he had fed Jonathan Lebed information about the companies, some of which turned out to be false, some of which Jonathan had unwittingly posted on Stock-dogs.com. The SEC asked if Monas had paid Jonathan to do this, and thus help to inflate the price of his company's stocks. Jonathan said no, he'd done it for free, because he'd thought the information was sound. Indeed, he had done it, at least at first, without any prompting from Monas. Monas had contacted him *after* he plugged Monas's companies. The SEC expressed its doubt that Jonathan was being forthright about his relationship with Monas. They had evidence that the man and the boy were thicker than the boy was admitting— or so they thought. One of the small companies Monas had been hired to plug was a cigar retail outlet called Havana Republic. As a publicity stunt Monas announced that the company—in which Jonathan came to own 100,000 shares— would hold a "smoke-out" in midtown Manhattan. The SEC now said that they knew that Jonathan Lebed had attended the smoke-out. To the people across the table from Jonathan this suggested that he had a normal face-to-face relationship with a known criminal:

SEC: *So you decided to go to the smoke-out?*

JONATHAN: *Yes.*

SEC: *How did you go about that?*

JONATHAN: *We walked down the street and took a bus.*

SEC: *Who is "we"?*

JONATHAN: *Me and my friend Chuck.*

SEC: *Okay.*

JONATHAN: *We took a bus to New York.*

SEC: *You cut school to do this?*

JONATHAN: *It was after school. Then we got held up at Port Authority, so then my mother and Harold came and picked us up and we went to the smoke-out.*

SEC: *Why were you picked up at the Port Authority?*

JONATHAN: *Because people like under 18 across the country, from California . . .*

MRS. LEBED: *They pick up minors there at Port Authority.*

SEC: *So the cops were curious about why you were there?*

JONATHAN: *Yes.*

SEC: *And they called your mother?*

JONATHAN: *Yes.*

SEC: *And she came.*

JONATHAN: *Yes.*

SEC: *You went to the smoke-out.*

JONATHAN: *Yes.*

SEC: *Did you see Ira there?*

JONATHAN: *Yes.*

SEC: *Did you introduce yourself to Ira?*

JONATHAN: *No.*

Here you can almost hear the little sucking sound on the SEC's side of the table, as the conviction goes out of this line of questioning.

SEC: *Why not?*

JONATHAN: *Because I'm not sure if he knew my age or anything like that, so I didn't talk to anybody there at all. I just wanted to go see everything with the company but I didn't want to associate with anybody there.*

SEC: *Did Ira Monas have any concept of your age?*

JONATHAN: *I don't know.*

SEC: *So you were thirteen at the time?*

JONATHAN: *Yes.*

SEC: *Did you ever discuss with Ira Monas that you won this stock contest [the CNBC contest]?*

JONATHAN: *No.*

SEC: *Do you have any reason to believe that he knows that?*

JONATHAN: *No.*

SEC: *Have you ever talked about that stock contest on the Internet?*

JONATHAN: *No.*

SEC: *Why is that?*

JONATHAN: *Because if I talked about it on the Internet people would just know that I'm fourteen and would just ignore me.*

This mad interrogation began at ten in the morning and ended at six in the evening. When they were done the SEC declined to offer legal advice. Instead they said, "We do wish to give our deep concern for the situation here, where unscrupulous people are perhaps using him in a way that he's not aware of . . . the internet is a grown-up medium for grown-up type of activities." Connie Lebed and Harold Burk, both clearly unnerved, apologized profusely on Jonathan's behalf, and explained that he was just a naïve child who had sought attention in the wrong place. Whatever Jonathan thought, he kept to himself.

"When I came home that day I closed the Ameritrade account," Connie now said.

"Then how did Jonathan continue to trade?" I asked.

The untamed end of the Lebed dining room bestirred itself; the Ape Drape gave a furious shake. "The kid never did something wrong!" Greg blurted out. He rose from his chair at the dining room table and paced back and forth beside the exotic-fish tank. This was how he got before he completely lost it and retreated to his trains in the basement. The trains, he said, were the only things that calmed him down.

"Don't ask me!" said Connie. "I got nothing to do with it."

"All right," said Greg, "youse wanna know what happened? Here's what happened. When Little Miss Nervous over here closes the Ameritrade account, I open an account for him in my name with that other place, E-Trade."

Thus did the impulse of the ordinary American working man to lift a middle finger in the general direction of his government find its way into the stock market. Greg had voted for Perot, and it showed. I turned to Jonathan, who wore his expression of airy indifference.

"But weren't you scared to trade again?"

"No."

"This thing with the SEC didn't even make you a little nervous?"

"No."

"No?"

"Why should it?"

Soon after he agreed to take Jonathan's case Kevin Marino discovered he had a problem. No matter how he tried, he was unable to get Jonathan Lebed to say what he really thought. "In a conversation with Jonathan I was supplying way too many of the ideas," Marino said. "You can't get them out of him." Finally he asked Jonathan and his parents each to write a few paragraphs describing their feelings about the SEC's treatment of Jonathan. Connie Lebed's statement took the form of a wailing lament about the pain inflicted by the callous government regulators on the family. ("I am also upset as you know that I was not called.") Greg Lebed's statement was an angry screed directed at both the government and the media. "We have a man in the Oval Office," he wrote, "who has disgraced the Presidential image of our whole country, but he continues to escape prosecution. My son Jonathan is touted as the first Stock Teen Swindler in this country and defends himself with his own monies. . . . This makes me sick!"

Jonathan's statement was so different in both tone and substance from his parents' that it inspired wonder that it could have been written by even the most casual acquaintance of the other two. It was on a Friday evening that Marino gave the family their assignment. Early Saturday morning he had a four-page e-mail from Jonathan. It began:

> I was going over some old press releases about different companies. The best performing stock in 1999 on the Nasdaq was Qualcomm [QCOM]. QCOM was up around 2000% for the year. On December 29th of last year, even after QCOM's run from 25 to 500, Paine Webber analyst Walter Piecky came out and issued a buy rating on QCOM with a target price of 1,000. QCOM finished the day up 156 to 662. There was nothing fundamentally that would make QCOM worth 1,000. There is no way that a company with sales under $4 billion, should be worth hundreds of billions. In the weeks to come, QCOM continued strong, going as high as 800 in early January. QCOM was being driven, not by fundamentals, but simply by the fact that a high price target was issued and many people were trading it. QCOM has now fallen from 800 to under 300. It is no longer the hot play with all of the attention. Many people were able to successfully time

QCOM and make a lot of money. The ones who had bad timing on QCOM, lost a lot of money.

People who trade stocks, trade based on what they feel will move and they can trade for profit. Nobody makes investment decisions based on reading financial filings. Whether a company is making millions or losing millions, it has no impact on the price of the stock. Whether it is analysts, brokers, advisors, Internet traders, or the companies, everybody is manipulating the market. If it wasn't for everybody manipulating the market, there wouldn't be a stock market at all. . . .

As it happens, those last two sentences stand for something like the opposite of the founding principle of the United States Securities and Exchange Commission. To a very great extent the rules and attitudes that govern not just the American but the world's financial markets are premised on a black-and-white mental snapshot of the American investor that was taken back in the early 1930s. And in the early 1930s the American investor was a caricature of panic and suspicion. The SEC was created in 1934, and the big question in 1934 was: How do you reassure the public that the stock market is not rigged? Between 1929 and 1932 the value of the stocks listed on the New York Stock Exchange had fallen 83 percent, from $90 billion to $16 billion. Capitalism, with reason, was not feeling terribly secure. To the greater public in 1934 the numbers on the stock market ticker no longer seemed to represent anything "real," but rather the end result of elaborate manipulation by financial pros. "I don't know, we don't know, whether values we have are going to be real next month or not," said one of the great stock market promoters of that age, Charles Schwab Sr.*

So: how to make the market seem "real"? The answer was to make stringent new laws against stock market manipulation. These were aimed not at ordinary Americans, who were assumed to be the potential victims of any manipulation, and the ones who needed to be persuaded that the market was not some elaborate, and ultimately empty, illusion. They were aimed at the Wall Street elite. The American financial elite acquired its own police force, whose job it was to make sure their machinations did not ever again unnerve the great sweaty rabble. That's not how the SEC put it, of course. The catchphrase used by the policy-making elites, when describing the SEC's mission, was "to restore public confidence in the securities markets." But it amounted to the same thing. Keep up appearances so that the public did not become too cautious. For the truth was that an excess of caution, by drying up the supply of risk capital, could doom capitalism.

It occurred to no one that the public might one day be as sophisticated in these matters as financial professionals. No one imagined the day when a fifteen-year-old boy might cause bedlam in the stock market, then write to his lawyer a withering dissection of that market:

A news release just came out today on SPACE.com, the Internet media venture founded

---

*For a detailed history of the SEC, and this quote, see Joel Seligman, *The Transformation of Wall Street*.

by former CNN Moneyline host Lou Dobbs. The company said that it would "lay off 22 workers, or 20 percent of its staff." They explained that this move is taking place, "in a push to turn profitable sooner than previously planned and achieve Dobbs' goal of diversifying beyond the Web."

We both know that the lay offs are being done not as a strategic move, but are taking place because the company is in a financial crisis and may not be around much longer. The comments about turning profitable sooner and diversifying beyond the Web just prove that the entire business and financial world revolves around public perception and nothing else but that.

That sounded cynical but really it wasn't, not exactly. Cynicism implies the possibility of idealism. Idealism about the financial markets simply would not occur to anyone who came of age in the second half of the 1990s.

Anyone who paid attention to the money culture could see its foundation had long lain exposed and that it was just a matter of time before the termites got to it. That had been one of the subplots of the Internet Boom. The signature trait of the boom was its from-the-ground-up quality. Little guys no one had ever heard of created companies worth billions of dollars more or less overnight. This was possible only because they had access to capital that, for little guys, was unprecedented. There had been a near-total collapse of the old financial order, in which a person, if he wished to borrow money, must already have money. The explosion in capital available for new ventures—and the new people who create them—was a natural extension of trends that were up and running by the early 1980s, when Michael Milken raised capital for all sorts of implausible characters (T. Boone Pickens, Nelson Peltz) who wanted to take over big corporations. The reason this change occurred was simple: it paid. It still pays.

The demise of Wall Street Man was just a curious consequence of the democratization of capital. First came the several billion dollars in ads from on-line brokers, a wave of anticapitalist propaganda unthinkable in any other period in American history, that sought to persuade the average American investor that the typical Wall Street stockbroker was a moron. Then the web sites arrived and suggested he was also a liar. From the moment the Internet went global back in 1996, web sites had popped up in the middle of nowhere—Jackson, Missouri, Carmel, California—and began to give away for free precisely what Wall Street sold for a living: earnings forecasts, stocks recommendations, market color. By the summer of 1998 Xerox or AT&T or some such opaque American corporation would announce earnings of 22 cents a share, and even though all of Wall Street had predicted a mere 20 cents and the company had exceeded their expectations, the stock would collapse. The amateur web sites had been saying 23 cents. In the fall of 1999 the Bloomberg news service commissioned a study to explore the phenomenon of what were now being called "whisper numbers." The study showed that the whisper numbers, the numbers put out by the amateur web sites, were mistaken, on average, by 21 percent. The professional

Wall Street forecasts were mistaken, on average, by 44 percent. The reason the amateurs now held the balance of power in the market was that they were, on average, more than twice as accurate as the pros—this in spite of the fact that the entire financial system was rigged in favor of the pros. The big companies spoon-fed their scoops directly to the pros; the amateurs were left in the dark.

The point was that even a fourteen-year-old boy could see how it all worked—why some guy working for free out of his basement in Jackson, Missouri, was more reliable than the most highly paid analyst on Wall Street. The companies that the financial pros were paid to analyze were also the financial pros' biggest customers. Xerox and AT&T and the rest needed to put the right spin on their quarterly earnings. The goal at the end of every quarter was for the newspapers and the cable television shows and the rest to announce that they had "exceeded analysts' expectations." The easiest way to exceed analysts' expectations was to have the analysts lower them. And that's just what they did—had been doing for years. The guy in Carmel, California, confessed to Bloomberg that all he had to do to improve on the earnings estimates of Wall Street analysts was to raise all of them 10 percent.

It didn't take long to realize what that meant—not just for the analysts but for the idea of the "financial professional," and indeed the long-term behavior of capital. The status of the financial professional was severely weakened, and capital would pay less attention to what he had to say the next time around. Of course, the professional financial analysts continued to insist that they were incorruptible—that there were "Chinese Walls" between their analysis and the bankers who actually collected the fees from the grateful corpora-tions. But it would have occurred to no one who was actu-ally investing via the Internet, and who could see the golden thread that links all common interests, to believe them. The system was rigged, the Internet had exposed the rigging, which meant that the system couldn't survive for long, at least not in its present form. Capitalism was eating the capitalists.

A few months after Bloomberg published its study the Internet bubble burst. The bust exposed the hollowness of the pros in a way that the boom never could. The market wiped out $5 trillion of value. The world hadn't experi-enced $5 trillion in change. A lot of people at once went looking for someone to blame. And the analysts went from looking ridiculous to people who watched them closely to looking ridiculous—and loathsome—to everyone. The most famous analysts on Wall Street, who had just a few weeks ago done whatever they could to cadge an appearance on CNBC, or a quote in the *Wall Street Journal*, to promote their favorite dot-com, went into hiding. Morgan Stanley's Mary Meeker, who had made $15 million in 1999 while telling people to buy Priceline when it was at $165 a share and Healtheon/WebMD when it reached $105 a share, went silent as they collapsed toward zero. What was the point of talking to the public if the talk wasn't going to drive the stock up? Merrill Lynch held encounter groups for its ana-lysts to teach them how to downgrade a company. The head of European stock market research at J. P. Morgan Chase, one Peter Houghton, sent around a memo to the bank's analysts. If the analysts planned to downgrade a company, it said, "advance copies of the complete research note must be sent to both parties [the company under analysis and the investment banker at J. P. Morgan Chase who did business with that company]. If the company requests changes to the research note, the analyst has a responsibility to either incorporate the changes requested or communicate clearly why the changes cannot be made."

Financial professionals now simply took it for granted that their job was to curry favor with the corporations about which they were meant to render honest judgments on behalf of the investing public. In the pursuit of banking fees the idea that there was such a thing as the truth had been lost.

To anyone who wandered into the money culture after, say, January 1996, it would have seemed absurd to take anything said by putative financial experts at face value. There was no reason to get worked up about this. Capitalism was safe for the rest of time. The stock market was not an abstraction whose integrity needed to be preserved for the sake of democracy. It was a game people played to make money. Who cared if anything anyone said or believed was "real"? The world could now afford to view money as no different from anything else that was bought or sold. Or, as Jonathan Lebed wrote to his lawyer:

> Every morning I watch Shop At Home, a show
> on cable television that sells such products as
> baseball cards, coins, and electronics. Don West,
> the host of the show, always says things like,
> "This is one of the best deals in the history of
> Shop At Home! This is a no brainer folks! This
> is absolutely unbelievable, congratulations to

everybody who got in on this! Folks, you got to get on the line, this is a gift, I just can't believe this!" There is absolutely nothing wrong with him making quotes such as those. As long as he isn't lying about the condition of a baseball card, or lying about how large a television is, he isn't committing any kind of a crime. The same thing applies to people who discuss stocks.

In the second half of the 1990s people like Jonathan Lebed, with the most innocent of motives, the desire to be rich, made a mockery of the financial order. This was maybe even more threatening than ordinary fraud to the bureaucratic self-interests of the SEC. The financial profession and its Washington regulators, though often at odds with each other, shared this view. Right from the start the SEC had helped to reinforce the sense that "high finance" was not something for ordinary people. It was conducted by elites. That the elites needed to be watched in a way that ordinary people did not merely further reinforced their status as elites.

Right from the start, the SEC treated the publicity surrounding the case of Jonathan Lebed at least as seriously as the case itself. Maybe even more seriously. The Philadelphia office had prosecuted the case, and so, when the producer from *60 Minutes* called to say he wanted to do a big piece about the world's first teenage stock market manipulator, he called the Philadelphia office. "Normally we call the top and get bumped down to some flak," said Trevor Nelson, the *60 Minutes* producer in question. "This time I left a message at the SEC's Philadelphia office and Arthur Levitt's office called me right back."

Levitt was the SEC's chairman—had been since 1993. He insisted on flying right up to New York City to be on the show. When he arrived he brought with him an intense concern about how the story would be told. Nelson recalled, "When Levitt came onto the set he asked me what I thought of the case and I remember having to think what the party line was before I said anything. Then Steve Kroft [the *60 Minutes* interviewer] shows up. Levitt asks him, 'How would you feel if this was your kid?' And Steve says, 'I'd be five hundred thousand richer. I'd feel pretty good about that.' Levitt turns white—and he's a pretty white guy. Then Don Hewitt [the legendary boss of *60 Minutes*] shows up. Levitt and Don know each other. You can see that Don hasn't thought about what the right answer was either, because Levitt turns to him and says, 'What do you think of what this kid did?' and Hewitt puts two thumbs up and booms: 'MORE POWER TO HIM.' Then he saw from the look on Levitt's face that that wasn't the right answer, and so he made a few concerned noises and started to leave. Before he could get out of the room Levitt said to him, 'Don, I hope you aren't going to make a hero of this kid.'"

To the SEC it wasn't enough that Jonathan Lebed hand over his winnings; he had to be vilified. There were victims out there! Innocent investors! "The SEC kept saying that they were going to give us the name of one of the kid's victims so we could interview him," says Nelson. "But they never did. I assume they went looking for a victim and couldn't find him."

I waited a couple of months for things to cool off before

heading down to Washington to see Arthur Levitt. He was just then finishing up being the second-longest-serving-ever chairman of the SEC and was taking a victory lap in the media for a job well done. He was now sixty-nine, but as a youth, back in the 1950s and 1960s, he'd made a lot of money on Wall Street, working for a firm whose dealings he later came to view as vaguely corrupt. Not illegal, corrupt. He'd do this thing with reporters who came to his office. Sotto voce, he'd explain how he had watched bankers and brokers conspire with company CEOs to focus on the price of a company's stock, rather than the business of the company. He didn't need to say that one of his former partners was Sandy Weill, the current chairman of Citigroup, the world's biggest financial firm. He'd leave it to others to draw the conclusions.

Levitt, at the age of sixty-two, had landed his job at the SEC—in part because he'd raised a lot of money on the Street for Bill Clinton. He arrived in Washington intent on ending his career more righteously than he had begun it. He had set himself up to defend the interests of the ordinary investor, and to a very great extent he'd succeeded. He'd declared war on the financial elite and pushed through new rules that stripped them of what was left of their old market advantages. His single bravest act was Regulation FD, which required corporations to release any significant information about themselves to everyone at once, rather than through the Wall Street analysts. The effect of this of course was to further empower traders like Jonathan Lebed, who could now be fairly certain that there wasn't some piece of information in the hands of the pros that they didn't know about.

I'd met Levitt only once, briefly, and had no idea what to

expect. He'd never asked why I wanted to see him—the meeting had been set up by one of his underlings, who hadn't asked either—and so I could only assume he had no idea. When he came to collect me in his waiting room—in the egalitarian spirit of a man who would prefer to dispense with the formalities of waiting rooms—he greeted me warmly. He knew my father, Joe Thomas, he said. Joe Thomas had been the head of the old Wall Street firm of Lehman Brothers long ago in its glory days. "He's not my father," I said. "Sure he is," said Levitt. You're thinking of Michael Thomas, I thought but did not say. Thomas was a journalist sympathetic to virtually any attempt by the SEC to defenestrate Wall Street financiers, exactly the sort of journalist who might be admitted into Levitt's presence, no questions asked. But the chairman was the sort of man who seems, at first blush, incapable of error. I even detected a note of uncertainty in my voice as I said, "But my last name isn't Thomas. It's Lewis."

For the next few seconds the chairman's face wore the expression of a man whose brain was undergoing the human equivalent of rebooting. Checking for viruses. He stared at me with deep and vacant eyes. I could see that this must be his meaningful stare. Before he could talk to me, it seemed, he had to place me. "What do you think of Charlie Wilson?" he finally asked, with a wicked little smile.

Which would seem a strange way to move the conversation forward. But actually it wasn't. Wilson—which wasn't his real name—was another journalist who, like Michael Thomas, could be counted on to celebrate any misery the SEC visited on Wall Street people. He was, from Arthur Levitt's point of view, safe. I told him that after Wilson reviewed my first book favorably he immediately hit me up for a blurb on his own forthcoming book, telling me point-blank that I would prove myself ungrateful if I did not return the favor. Back then I thought this sort of thing was corrupt—back then I didn't know how the book business worked—and so I hadn't done it. Still, that told me something about Charlie Wilson: he was just another opportunist masquerading as a moralist, and I was weary of pretending to admire people like that.

Levitt's disappointment that we wouldn't be sharing a conspiratorial chuckle over the hell visited by muckrakers on Wall Street big shots seemed to be offset by his satisfaction in the fact that I appeared to be aflame with high principle. I, too, was safe. Comforted, he set out to explain to me the new forces corrupting the financial markets. "The Internet has speeded up everything," he said, "and we're seeing more people in the markets who shouldn't be there. There's what I like to call the mythology of the trader. A lot of these new investors don't have the experience, or the resources, of a professional trader. These are the ones who bought that crap that Lebed was pushing."

"Do you think he is a sign of a bigger problem?" I asked.

"Yes, I do. And I find his case very disturbing . . . more serious than the guy who holds up the candy store—because it says something about values. I think there's a considerable risk of an antibusiness backlash in this country. The era of the twenty-five-year-old billionaire represents a kind of symbol which is different from the Horatio Alger symbol. The twenty-five-year-old billionaire looks lucky, feels lucky. And investors who lose money buying stock in the company of the twenty-five-year-old billionaire . . ."

He trailed off, leaving me to finish the thought.

"You think it's a moral issue."

"I do. I think there is a values problem."

"You think Jonathan Lebed is a bad kid?"

"Yes, I do. . . . That Mercedes he bought. That is the part of the case I found very, very disturbing."

"Can you explain to me what he did?"

Another meaningful stare; his eyes were light blue bottomless pits. "He'd go into these chat rooms and use twenty fictitious names and post messages. . . ."

"By fictitious names do you mean e-mail addresses?"

"I don't know the details."

Don't know the details? He'd been all over the airwaves decrying the behavior of Jonathan Lebed.

"Put it this way," he said. "He'd *buy, lie, and sell high.*" The chairman's voice had deepened unnaturally. He hadn't spoken the line; he'd acted it. It was exactly the same line he had spoken on *60 Minutes* when his interviewer asked him to explain Jonathan Lebed's crime. He must have caught me gaping in wonder because he shifted and for the third time looked at me long and hard. I glanced away.

"What do you think?" he asked.

Well, I had my opinions. In the first place, I had been surprised to learn that it was legal for, say, an author to write phony glowing reviews of his latest book on Amazon.com but illegal to plug a stock on Yahoo Finance message boards just because he happened to own it. I thought it was naïve to pretend that the sort of people who read Crazy Eddie–style promotions on Yahoo Finance message boards and went out and bought the stock were innocent victims when anyone who spent more than an hour on Yahoo could see they were surfers hoping to catch a wave, and as sophisticated about the fluctuations of the stock market as any SEC chairman. I also thought it was—to put it kindly—misleading to tell reporters that Jonathan Lebed had used "fictitious names" when he had used four AOL e-mail addresses, and posted exactly the same message under each of them, so that no one who read them could possibly mistake him for more than one person. I further thought that, without quite realizing what had happened to them, the people at the SEC were now lighting out after the very people—the average American with a bit of money to play with—that they were meant to protect.

Finally I thought that by talking to me or any other journalist about Jonathan Lebed when he didn't really understand himself what Jonathan Lebed had done, the chairman of the SEC displayed a disturbing faith in the media to buy whatever he was selling. He'd grown a bit too used to being cast as the hero of the story.

But when he asked me what I thought, all I said was, "I think it's more complicated than you think."

"Richard! Call Richard!" Levitt was shouting out the door of his vast office. "Tell Richard to come in here!"

Richard was Richard Walker, the SEC's director of enforcement. He entered with a smile but mislaid it before he even sat down. His mind went from a standing start to deeply distressed inside of ten seconds. "This kid was making predictions about the prices of stocks," he said testily. "He had no basis for making these predictions." Before I can tell him that that sounds a lot like what happens every day on Wall Street, he says, "And don't tell me that's standard practice on Wall Street," so I didn't. But it was, and still is. It is okay for the analysts to lowball their estimates of corporate earnings so that they remain in the good graces of those companies. It was okay for analysts to plug companies with one hand and collect fees from them with the other. The SEC might protest that the analysts don't actually *own* the stocks they plug, but that is a distinction without a moral difference: they profit mightily from the stock's rise. It was okay that Mary Meeker of Morgan Stanley and Henry Blodget of Merrill Lynch had plugged a portfolio of Internet company shares that, inside of six months, lost more than three quarters of their value at the same time that they were paid millions of dollars, largely as a result of the fees their firms raked in from the very same Internet companies. But it was, for some reason I do not fully grasp, not okay for Jonathan Lebed to say that FTEC would go from 8 to 20. When I asked *why* it was illegal, Walker had a pat answer.

"Because Jonathan Lebed was seeking to *manipulate* the market," he said.

But that only begged the question. If Wall Street analysts and fund managers and corporate CEOs who appear on CNBC and CNNfn to plug stocks are not guilty of seeking to *manipulate* the market, what on earth does it mean to *manipulate* the market?

"It's when you promote a stock for the purpose of artificially raising its price," said Walker.

But when a Wall Street analyst can send the price of a stock of a company that is losing billions of dollars up fifty points in a day, what does it mean to "artificially raise" the price of a stock? The law sounded perfectly circular. Actually, this point had been well made in a recent article in *Business Crimes Bulletin* written by a pair of securities law experts, Lawrence S. Bader and Daniel B. Kosove. "The case books are filled with opinions that describe manipulation as causing an 'artificial' price," the experts wrote. "Unfortunately, the case books are short on opinions defining the word 'artificial' in this context. . . . By using the word 'artificial' the courts have avoided coming to grips with the problem of defining 'manipulation'; they have simply substituted one undefined term for another."

"The price of a stock is artificially raised when subjected to something other than ordinary market forces," Walker recites.

"But what are 'ordinary market forces'?"

An ordinary market force, it turned out, was one that does not cause the stock to rise artificially. In short, an ordinary market force is whatever the SEC says it is, or what it can persuade the courts it is. And the SEC does not view teenagers broadcasting their opinions as "an ordinary market force." It can't. If it did, it would be compelled to face the deep complexity of the modern market—and all of the strange new creatures who have become, with the help of the Internet, ordinary market forces. That's what happened when the Internet collided with the stock market: Jonathan Lebed became a market force. *Adolescence* became a market force.

I finally came clean with my suspicion that there was a reason the SEC let Jonathan Lebed walk away with five hundred grand in his pocket—even though its initial press release suggested it had forced him to repay all his profits. The people who worked at the SEC feared that if they didn't cave, they'd wind up in court, and they'd lose. And if the law ever declared formally that Jonathan Lebed didn't break it, the SEC would be faced with an impossible situation: millions of small investors plugging their portfolios with abandon, becoming, in essence, professional financial analysts, generating little explosions of unreality in every corner of the capital markets. No central authority could sustain the illusion that stock prices were somehow "real," or that the market wasn't, for most people, a site of not terribly productive leisure activity. The red dog would be off his leash.

A large part of what the U.S. Securities and Exchange Commission does for a living is attempt to stamp out what it calls "market manipulation." In saying that market manipulation is the rule and not the exception in the markets, I crossed some invisible line inside the SEC chairman's office. I might as well have strolled up to the drug czar and lit a joint.

"The kid himself said he set out to manipulate the market!" Walker virtually shrieked. But of course that is not all the kid said. The kid said *everybody* in the market was out to manipulate the market.

"Then why did you let him keep five hundred grand of his profits?" I asked.

"We determined that those profits were different from the profits he made on the eleven trades we defined as illegal," he said.

This, I already knew, was a pleasant fiction. The amount Jonathan Lebed handed over to the government was determined by haggling between Marino and the SEC's Philadelphia office. The SEC initially demanded the $800,000 Jonathan had made, plus interest. Marino had countered at

125 grand. The SEC recountered at 350 grand. They hag-
gled a bit more and then settled at 285.

"Can you explain how you distinguished the illegal
trades from the legal ones?" I asked.

"I'm not going to go through the case point by point," he
said.

"Why not?"

"It wouldn't be appropriate."

At which point Arthur Levitt, who had been attempting
to stare into my eyes as intently as a man can stare, said, in
his deep voice, "This kid has no basis for making these predictions."

"But how do you *know* that?"

And the chairman of the SEC, the embodiment of
investor confidence, the keeper of the notion that the numbers gyrating at the bottom of the CNBC screen are "real,"
draws himself up and says, "I *worked* on Wall Street."

Well. What do you say to that? He had indeed worked on
Wall Street—in 1968.

"So did I," I said.

"I worked there longer than you."

Walker leapt back in. "This kid's father said he was going
to rip the fucking computer out of the wall."

Probably that much was true. The second time the SEC
called on the Lebed household they began with Greg. This
time they meant business. They hauled Greg down to the
Philadelphia office and questioned him for six hours. This,
to judge from the transcript, made for another bizarre
scene. A row of SEC lawyers on one side of the table; one
seriously pissed-off guy from the tough side of New Jersey on
the other. Waco Meets Wall Street. Somehow the SEC's

investigators, when they were dealing with the Lebeds from their usual distance, hadn't been able to get their mind around the possibility that the evil genius they were after was a fifteen-year-old boy. But once they had Greg in the flesh it took the lawyers about three minutes to reconsider their prejudice.

> SEC: *Did he tell you what he was doing?*
> GREG: *No, not exactly; no.*
> SEC: *Were you interested in finding out what he was doing on the Internet which would have caused the SEC to come looking for him?*
> GREG: Well . . . the wife went and found out what that was all about.

The wife.

> SEC: *Right, but my question to you, sir, is whether you had enough concern about why the SEC needed to talk to your son, who was then 14 years old, did you have enough concern that you wanted to find out what he was doing on the Internet?*
> GREG: *I think I might have hollered at him for getting involved with something he didn't know nothing about, like this guy that could be misleading him or a crook. I approached it from that angle like you are not supposed to be doing something for somebody or don't listen to somebody that you don't know who the person is from Adam.*
> SEC: *So you were yelling at Jonathan but you don't know what he was doing? . . . Do you think that—in your*

> *opinion—in the approximately 12 months you gener-*
> *ate 800 thousand dollars in profits, is that unusual?*
>
> GREG: *I think it is highly probable in this day and age*
> *with this Internet stuff.*
>
> SEC: *Do you think it is highly probable for a 15 year old*
> *to do that?*
>
> GREG: *He is very intelligent.*

The SEC then asked Greg the obvious question to pose to the father of a boy dragged in by the U.S. government not once but twice. It was the same question I had asked: How could he possibly have permitted his son to keep trading *after* the first dustup?

> GREG: *Well, he still wanted to buy stock, so I said to*
> *him, "Look, if you want to buy stocks, we are going to*
> *open the account in my name but whatever you do, I*
> *don't want you getting into trouble with no Internet*
> *bullshit like you did last year." He said, "Okay, dad."*
> *Now during the year, different times, I have always*
> *questioned him, "How is everything going? No prob-*
> *lems?" He said, "No," that everything was okay. And*
> *that was it.*
>
> SEC: *If Jonathan was on the Internet doing the exact*
> *same thing he got in trouble for with the SEC in Octo-*
> *ber, how would you know that he was doing it if he*
> *didn't tell you?*
>
> GREG: *I wouldn't or let's say I didn't if he is.*
>
> SEC: *So are you saying that basically after the contact*
> *that Jonathan had with the SEC in October, you relied*

> upon Jonathan's word as to whether or not he was
> doing it?
>
> GREG: *Yes, I did.*
>
> SEC: *So you relied on him to determine whether he was
> staying out of trouble essentially?*
>
> GREG: *I wouldn't—how would you know if a kid is on
> the computer like 10 million are every day, how do you
> monitor it?*

There it was again: the strange inversion of authority afflicting the Lebed family. You might have thought that the SEC would come to terms with the fact that they were now attempting to police a nation of native-born immigrants. On the other hand, it's not the SEC's job to figure out what financial life is actually like out there in America. Their job is to insist on the validity of the old black-and-white snapshot.

More silence than is healthy had descended upon Arthur Levitt's office. I realized that it was my turn to stare. It was a waste of time trying to stare at Arthur Levitt— like trying to outstare a black hole—so I stared at Richard Walker. "Have you *met* Jonathan Lebed's father?" I said.

"No, I haven't," he said, curtly. "But look, we talked to this kid two years ago, when he was *fourteen years old.* If I'm a kid, and I'm pulled in by some scary government agency, I'd *back off.*"

That's the trouble with fourteen-year-old boys—from the point of view of the social order. They haven't yet learned the more sophisticated forms of dishonesty. It can take years of slogging to learn how to feign respect for hollow authority.

Still! That a fourteen-year-old boy, operating essentially in a vacuum, would walk away from a severe grilling by six hostile bureaucrats who commanded the resources of the most powerful government on earth and jump right back into the market: how did *that* happen? It occurred to me, as it had occurred to Jonathan's lawyer, that I had taken entirely the wrong approach to getting the answer. The whole point of Jonathan Lebed was that he had invented himself on the Internet. The Internet had taught him how hazy the line was between perception and reality. When people could see him, they treated him as they would treat a fourteen-year-old boy. When all they saw were his thoughts on financial matters, they treated him as if he were a serious trader. On the Internet, where no one could see who he was, he became who he was.

I left the SEC and went back to my hotel and sent Jonathan an e-mail. I asked him the same question I had asked the first time we had met: Why hadn't he been scared off? Straightaway he wrote back:

> It was about 2–3 months from when the SEC called me in for the first time until I started trading again. The reason I didn't trade for those 2–3 months is because I had all of my money tied up in a stock. I sold it at the end of the year to take a tax loss which allowed me to start trading again. I wasn't frightened by them because it

There were no explicit rules on Yahoo, but there were constraints. The first was that Yahoo limited the number of messages Jonathan could post using one e-mail address. He'd click onto Yahoo and open an account with one of his four AOL screen names; a few minutes later Yahoo, mysteriously, would tell him that his messages could no longer be delivered. Eventually he figured out that they must have some limit that they weren't telling people about. He got around it by grabbing another of his four AOL screen names and creating another Yahoo account. By rotating his four AOL screen names, he found he could get his message onto maybe two hundred Yahoo message boards before school.

He also found that when he went to do it the next time, with a different stock, Yahoo would no longer accept messages from his AOL screen names. So he was forced to create four more screen names and start over again. Yahoo never told him he shouldn't do this. "The account would be just, like, deleted," he says. "Yahoo never had a policy; it's just what I figured out." The SEC accused Jonathan of attempting to seem like more than one person when he promoted his stocks, but when you see how and why he did what he did, that is clearly false. (For instance, he ignored the feature on Yahoo that enables users to employ up to seven different "aliases" for each e-mail address.) It's more true to say that he was attempting to simulate an appearance on CNBC.

At any rate, through much trial and error, Jonathan learned that some messages had more effect on the stock market than others. "I definitely refined it," he said of his Internet persona. "In the beginning I would write, like, very professionally. But then I started putting stuff in caps and

using exclamation points and making it sound more exciting. That worked better. When it's more exciting it draws
people's attention to it, compared to when you write, like,
dull or something." The trick was to find a stock that *he*
could get excited about. He'd stopped watching CNBC—
"There wasn't any good information on it and I don't think
any of the discussion is really very useful." Most of his dis-
coveries he made by poking around the Willow Brook Shopping Mall, down the road in Wayne, or on the Internet.
When he wasn't either asleep or in some dreary classroom,
he lived on the Internet, which he scoured for small cap
stocks that he thought might appeal to other Internet
investors. He sifted the small cap stocks with three things in
mind: (1) "It had to be in the area of the stock market that
is likely to become a popular play"; (2) "It had to be undervalued compared to similar companies"; and (3) "It had to
be undiscovered—not that many people talking about it on
the message boards."

All of which is to say that he didn't think of himself as
some Internet cowboy, but as a stock market investor. "I tried
to be very conservative," he explained. "It was less risky than
long-term investing. You have no idea what is going to happen in the long term. But if you buy something in the short
term there's always a good chance you'll be making money
on it sometime. I'd buy small positions and average down if
I had to. I'm not a day trader. Day traders usually have no
idea about the companies, and I have that knowledge. I tried
to think longer term, even though I was trading short term."

Here the boy once again was simply aping one of the
more curious developments in the modern financial mar-

kets. Over the past couple of decades the market has blurred the distinction between speculating and investing. Historically, this distinction hasn't been merely financial but social. Investors like Warren Buffett are widely regarded to be the salt of the earth, the rock on which the church is founded, the pillar of stability in this sea of decadence. Speculators like George Soros once were vaguely disreputable. But gambling has lost its status as a sin; and for some time now the world's smartest money has been pouring into hedge funds and venture capital funds, in search of quick killings. Highly speculative "investing" has become respectable.

But there was something else going on, too. Capital was coming to be treated as just another commodity, instead of some sacred trust. The people who enjoyed access to it were no longer viewed as members of an exclusive club, since there was a new feeling in the air that everyone, in theory, enjoyed access to it, including fifteen-year-old middle-class boys in Cedar Grove, New Jersey. What they did with capital, once capital came without taboos, was hardly surprising.

From Jonathan Lebed's desultory point of view, the behavior of the grown-ups who disapproved of what he had done with his money was hysterical, in both senses of the word. One day he picked up a newspaper and read an SEC investigator named Ron Long explaining to reporters that "at about 14, he [Jonathan] crossed into the dark side. While he was sitting in math and sciences, he knew he was making his profits." He turned on the television and heard that Richard Walker, the SEC's director of enforcement, had just given a speech to a bunch of Wall Street bond traders in which he'd said, "Maybe the Lebeds should have used some of Jonathan's profits to buy a dictionary rather than a Mercedes." (What did the bond traders drive?) Jonathan turned on the radio to hear Arthur Levitt say, "If he were my son, I would not be proud of him."

Who were these people? The astonishing truth was that Jonathan had never laid eyes on any of them. He'd never described to a single adult exactly what he had done on the Internet. So far as he could tell, the people at the SEC didn't really understand what he had done. No one did—for the simple reason that no one had ever asked him to show them. No one. Not his parents, not his friends, not his teachers, not the SEC: no one had *ever* asked him to sit down at the computer and to explain exactly what he had done or why he had done it. And so the only lesson Jonathan learned from his dealings with the SEC was that other grown-ups, when threatened, could become as madly enraged as his father.

Over a couple of months I drifted in and out of Jonathan Lebed's life and became used to its staccato rhythms. His mind was a peculiar combination of grandiosity and myopia. When he stared into his computer screen he saw the depths of the universe; when he looked up from it he was hardly able to see beyond Cedar Grove. From the moment he became a celebrity in town, he attended meetings of the Town Council, where he'd become a kind of freelance civic activist. He could imagine entering politics but couldn't imagine much beyond mayor of Cedar Grove, New Jersey. When I asked him where he was thinking about applying to college, he couldn't think much past Montclair, New Jersey. He had no more ability than I had to link his Internet self with his real-world one. He could imagine himself doing great things, but these great things always took place within a few hundred yards of where he happened to be standing—except when he went on the Internet.

His defining trait was that the strangest things happened to him and he just thought of them as perfectly normal—and there was no one around to clarify matters. He had a crowd of friends at Cedar Grove High School, most of whom owned pieces of Internet businesses and all of whom speculated in the stock market. "There are three groups of kids in our school," one of them explained to me. "There's the jocks, there's the druggies, and there's us—the more business-oriented. The jocks and the druggies respect what we do. At first a lot of the kids are, like, what are you doing? But once kids see money they get excited."

The first time I heard this version of the social structure of Cedar Grove High, I hadn't taken it seriously. But then one day I went out with Jonathan and one of his friends, Keith Graham, into a neighboring suburb to do what they liked to do most when they weren't doing business: shoot pool. We parked the car and set off down an unprosperous street in search of the pool hall. We made a strange sight: two kids in blue jeans and T-shirts and expressions of permanent boredom accompanied by one grown man attempting to scribble down their stray, infrequent remarks.

"Remember West Coast Video?" Keith said, drolly.

I looked up: we were walking past a derelict building with WEST COAST VIDEO stenciled on its plate glass. West

Coast Video had been one of several attempts to unseat Blockbuster. It failed, but before it did, put on a good show.

Jonathan chuckled knowingly. "We owned like half the company."

I looked at him: he was perfectly serious. He began ticking off the reasons for his investment. "First, they were about to open an Internet subsidiary; second, they were going to sell DVDs when no other video chain—"

I stopped him before he really got going. "*Who* owned half the company?"

"Me and a few others. Keith, Michael, Tom, Dan."

"Some teachers too," said Keith.

"Yeah, the teachers heard about it," said Jonathan. He must have seen me looking strangely at him because he added, "It wasn't that big a deal. We probably didn't have a controlling interest in the company. Just a fairly good percentage of the stock."

"I don't know," said Keith. "Tom and Dan had like fifteen, twenty grand in it. Dan had, like, taken control of his dad's retirement fund."

"*Teachers?*" I said. "The teachers followed you into this sort of thing?"

"Sometimes," said Jonathan.

"*All* the time," said Keith. Keith is a year older than Jonathan and tended to be a more straightforward narrator of events. Jonathan habitually dramatized or understated his case, and when he did, emitted a strange frequency, like a boy not quite sure how hard to blow into his new tuba. Keith invariably corrected him. "As soon as people at school found out what Jonathan was in, *everybody got in*," he said. "Like right way. It was, like, if Jonathan's in on it, it must be good. "

And with that the two boys moved on to some other subject, bored with the memory of having led their teachers in the acquisition of a meaningful chunk of the outstanding shares of West Coast Video. We entered the pool hall and took a table, where we were soon joined by another friend, John. Keith had paged him.

My role in Jonathan Lebed's life suddenly grew clear: to express sufficient wonder at whatever he got up to that he was compelled to elaborate. "I don't understand," I said. "How would other kids find out what Jonathan was in?"

"It's high school," said Keith, in a tone reserved for people over thirty-five. "Four hundred kids. People talk."

"How would the teachers find out?"

Now Keith shot me a look that tells me that I'm the most prominent citizen of a new nation called STUPID. "They would ask us!" he said.

"But why?"

"They saw we were making money," said Keith. "Money talks, bullshit walks."

"Yeah," said Jonathan, who, odd as it sounds, exhibited none of his friend's knowingness. He just *knew*. "I feel, like, that most of my classes, my grades would depend not on my performance but on how the stocks were doing."

"Not really," said Keith.

"Okay," said Jonathan. "Maybe not that. But, like, I didn't think it mattered if I was late for class."

Keith considered that. "That's true," he said.

"I mean," said Jonathan, "they were making like thousands of dollars on the trades, more than their salaries even."

"Look," I said, "I know this is a stupid question. But was there any teacher who . . . say . . . disapproved of what you were doing?"

The three boys considered this, plainly for the first time in their lives.

"The librarian," Jonathan finally said.

"Yeah," said John. "But that only because the computers were in the library and she didn't like us using them."

"You traded stocks from the library?"

"Fifth-period study hall was in the library," said Keith. "Fifth-period study hall was like a little Wall Street. But sometimes the librarian would say the computers were for study purposes only. None of the other teachers cared."

"They were all trading," said Jonathan.

The mood shifted. For the next few minutes we shot pool and pretended that there was no more boring place to be than this planet we live on. "Even though we owned like a million shares," said Jonathan, picking up the new mood, "it wasn't that big a deal. West Coast Video was trading at like thirty cents a share when we got in."

Keith looked up from the cue ball. "When *you* got in," he said. "Everyone else got in at sixty-five cents, then it collapsed. Most of the people lost money on that one."

"Hmmm," said Jonathan, with real satisfaction. "That's when I got out."

I realized then that the SEC had been right: there *were* victims to be found from Jonathan Lebed's life on the Internet. They were right here in Cedar Grove, New Jersey. I turned to Keith. "You're Jonathan's victim."

"Yeah, Keith," said Jonathan, laughing. "You're my victim."

"Nah," said Keith. "In the stock market you go in knowing you can lose. We were just doing what Jon was doing but not doing as good a job at it."

TWO

PYRAMIDS AND

PANCAKES

The strange story of Jonathan Lebed suggested that you couldn't really understand what was happening on the Internet unless you understood the conditions in the real world that led to what was happening on the Internet, and you couldn't understand those unless you went there in person and had a look around. And so it wasn't long after I left Jonathan that I found myself back on the road, heading south from Los Angeles into the desert, to investigate another unintentional insult delivered to the social order by a teenage boy with an AOL account. This one had occurred on a web site maintained by the AskMe Corporation.

The AskMe Corporation had been created in 1999 by former Microsoft employees. The software it sold enabled the big companies that had bought it—3M, Procter & Gamble—to create a private web for their workers. This private web was known as a "knowledge exchange." The knowledge exchange was a screen on a computer where employees could put questions to the entire company. The appeal of this was obvious. Once an AskMe-style knowledge exchange was up and running, it didn't matter where inside the company any particular expertise resided. So long as expertise didn't leave the company, it was always on tap for whoever needed it.

AskMe Corp. soon found that it was able to tell a lot about a company from its approach to the new software. In pyramid-shaped, hierarchical organizations, the bosses tended to appoint themselves or a few select subordinates as the "experts." Questions rose up from the bottom of the organization, the answers flowed down from the top, and the original hierarchy was preserved, even reinforced. In less hierarchical, pancake-shaped companies, the bosses used the software to create a network of all the company's employees and to tap intelligence wherever in that network it happened to be. That way anyone in the company could answer anyone else's questions. Anyone could be the expert. Of course, it didn't exactly inspire awe in the ranks to see the intern answering a question posed by the vice president of strategic planning. But many companies decided that a bit of flattening was a small price to pay to tap into the collective knowledge bank.

The people who created the AskMe.com software believed that it gave companies whose bosses were willing to risk their own prestige and authority an advantage over the hierarchical companies whose bosses were not. They didn't say this publicly, because they wanted to sell their software to the pyramid-shaped organizations, too. But they knew that once the software was deployed, companies that flattened their organization charts to encourage knowledge to flow freely in every direction would beat companies that didn't. Knowledge came from the strangest places; employees knew a lot more than they thought they did; and the gains in the collective wisdom outweighed any losses to the boss's authority.

In short, the software subtly changed the economic environment. It bestowed new rewards on the egalitarian spirit. It made life harder for pyramids and easier for pancakes.

Out in the field AskMe Corp.'s salespeople, like salespeople everywhere, found themselves running into the same five or six objections from potential buyers—even when the buyers were pancake-shaped. One was "How do you know that your software won't break down when all of our two hundred thousand employees are using it heavily?" To prove that it wouldn't, AskMe Corp. created a web site and offered a version of its software to the wider public. The site, called AskMe.com, went up on the Web in February 2000, and quickly became the most heavily used of a dozen or so knowledge exchanges on the Internet. In its first year the site had more than ten million visitors, which was striking in view of how peripheral it was to the ambitions of the AskMe Corporation. The company made no money from the site and did not bother to monitor what went on there, or even to advertise its existence. The ten million people who had used the site in its first year were drawn by word of mouth. The advice on the site was freely offered. The experts were self-appointed and ranked by the people who sought the advice. Experts with high rankings received small cash prizes from AskMe.com. The prizes—and the free publicity— attracted a lot of people who don't normally work for free. Accountants, lawyers, and financial consultants mingled their licensed knowledge with experts in sports trivia, fortune telling, and body piercing.

AskMe Corp didn't think of it this way, but its public web site suggested a question: What is the wider society's instinctive attitude toward knowledge? Are we willing to look for it wherever it might be found, or only from the people who are supposed to possess it? Does the world want to be a pyramid or a pancake?

In the summer of 2000, in a desert town called Perris, halfway between L.A. and Palm Springs, a fifteen-year-old boy had offered his reply to that question, and a thousand or so more besides. His name was Marcus Arnold. His parents had immigrated to Perris from Belize, by way of South-Central Los Angeles. Why anyone would move to Perris from anywhere was not immediately clear. Perris was one of those non-places that America specializes in creating. One day it was a flat hazy stretch of sand and white rock beneath an endless blue sky; the next some developer had laid out a tract of twenty-five thousand identical homes; and the day after that it was teeming with people who were there, main-ly, because it was not someplace else. The decision of human beings to make a home of it had little effect on the identity of Perris. Even after the tract houses had been deposited on the desert, Perris was known chiefly as a place to leap onto from an airplane.

Marcus lived with his parents and his twin brother in a small brick house a mile or so from the big drop zone. Over the family's one-car garage, from morning until night, peo-ple stepped out of planes and plummeted to earth, and the blue sky above Marcus was permanently scarred by para-chutes. Marcus himself was firmly earthbound, a great big bear of a boy. He was six feet tall and weighed maybe two hundred pounds. He did not walk but lumbered from the computer to the front door, then back again. The computer squatted on a faux antique desk in the alcove between the dining room and the living room, which were as immacu-lately kept as showrooms in a model home. It was the only computer in the house, he said. In theory, the family shared it; in practice, it belonged to him. He now needed as much time on it as he could get, as he was a leading expert on AskMe.com. His field was the law.

The blue screen displayed the beginning of an answer to a question on AskMe.com Marcus had bashed out before I arrived:

> Your son should not be in jail or on trial. According to *Miranda versus Arizona* the person to be arrested must be read his rights before he was asked any questions. If your son was asked any questions before the reading of his rights he should not be in prison. If you want me to help you further write me back on this board privately.

The keyboard vanished beneath Marcus's jumbo hands and another page on AskMe.com popped up on the screen. Marcus wanted to show me the appallingly weak answer to a question that had been offered by one of the real lawyers on the site. "I can always spot a crummy attorney," he said. "There are people on the web site who have no clue what they're talking about, they are just there to get rankings and to sell their services and to get paid." Down went his paws, out of sight went the keyboard, and up popped one of Marcus's favorite web sites. This one listed the menus on death row in Texas. Photographs of men put to death by the state appeared next to hideous lists of the junk food they'd ordered for their last meals. Marcus browsed these for a minute or two, searching for news, then moved on, without comment. One of the privileges of adolescence is that you can treat everything around you as normal, because you have nothing to compare it to, and Marcus appeared to be taking full advantage of it. To Marcus it was normal that you could punch a few buttons into a machine and read what a man who was executed by the state this morning had eaten last night. It was normal that the only sign of life outside his house were the people floating down from the sky and into the field out back. It was normal that his parents had named his identical twin brother Marc. Marc and Marcus. And it was normal that he now spent most of the time he was not in school on the Internet, giving legal advice to grown-ups.

Marcus had stumbled upon Askme.com late in the spring of 2000. He was studying for his biology exam and looking for an answer to a question. He noticed that someone had asked a question about the law to which he knew the answer. Then another. A thought occurred: why not answer them himself? To become an official expert he only needed to fill in a form, which asked him, among other things, his age. He did this on June 5—a day already enshrined in Marcus's mind. "I always wanted to be an attorney since I was, like, twelve," he said, "but I couldn't do it because everyone is going to be, 'like, what? Some twelve-year-old kid is going to give me legal advice?'"

"They'd feel happier with a fifteen-year-old?"

He drew a deep breath and made a face that indicated that he took this to be a complicated question. "So when I first went on AskMe," he said, "I told everybody I was twenty, roughly about twenty, and everyone believed me." Actually, he claimed to be twenty-five, which to a boy of fifteen is, I suppose, roughly twenty. To further that impression he adopted the handle "LawGuy1975." People who clicked onto his page found him described as "LawGuy1975 aka Billy Sheridan." "Billy Sheridan" was Marcus's handle on America OnLine.

A few days after he appointed himself a legal expert Marcus was logging onto the Internet solely to go to AskMe.com and deal with grown-ups' legal problems. What sort of legal problems? I asked him. "Simple ones," he said. "Some of them are like, 'My husband is in jail for murder and he didn't do it and I need to file a motion for dismissal, how do I do it?' I have received questions from people who are just, like, you know, 'I am going to be put in jail all of a sudden, can somebody help me plead before they come cart me off,' and it's just, like, well, come on, that's a cry for help. You're not just going to sit there. . . . But most of them are simple questions. 'What's a felony?' Or 'How many years will I get if I commit this crime?' Or 'What happens if I get sued?' Or 'What's the process to file papers?' Simple questions." He said all this in the self-conscious rapid-fire patter of a television lawyer.

Once he became an expert, Marcus's career took on a life of its own. The AskMe rankings were driven by the number of questions the expert answered, the speed of his replies, and the quality of those replies, as judged by the recipients, who bestowed on them a rating of between one and five stars. By July 1 Marcus was ranked number 10 out of one hundred and fifty or so experts in AskMe.com's criminal law division, many of whom were actual lawyers. As he tells it, that's when he decided to go for the gold. "When I hit the top ten I got some people who were like, 'Congratulations, blah blah blah.' So my adrenaline was pumping to answer more questions. I was just, like, 'You know what, let me show these people I know what I'm doing.'" He needed to inspire even more people to ask him questions, and to reply to them quickly, and in a way that prompted them to reward him with lots of stars. To that end he updated the page that advertised his services. When he was done it said:

I AM A LAW EXPERT WITH TWO YEARS OF FORMAL TRAINING IN THE LAW. I WILL HELP ANYONE I CAN! I HAVE BEEN INVOLVED IN TRIALS, LEGAL STUDIES AND CERTAIN FORMS OF JURISPRUDENCE. I AM NOT ACCREDITED BY THE STATE BAR ASSOCIATION YET TO PRACTICE LAW. . . . SINCERELY, JUSTIN ANTHO-NY WYRICK, JR.

"Justin was the name I always wanted—besides mine," Marcus said. Justin Anthony Wyrick Jr.—a pseudonym on top of a pseudonym on top of a pseudonym. Justin Anthony Wyrick Jr. had a more authoritative ring to it, in Marcus's opinion, and in a lot of other people's, too. On one day Marcus received and answered 110 questions. Maybe a third of them came from the idly curious, a third from people who were already in some kind of legal trouble, and the final third from people who appeared to be engaged in some sort of odd cost-benefit analysis.

> **Q:** What amount of money must a person steal or gain through fraud before it is considered a felony in Illinois?
> **A:** In Illinois you must have gained $5001+ in an illegal fashion in order to constitute fraud. If

you need anything else please write back! Sincerely, Justin Anthony Wyrick Jr.

Q: Can a parole officer prevent a parolee from marrying?

A: Hey! Unless the parolee has "No Marriage" under the special conditions in which he is released, he can marry. If you have any questions, please write back. Sincerely, Justin Anthony Wyrick Jr.

The more questions Marcus answered, the more people who logged onto the boards looking for legal advice wanted to speak only to him. In one two-week stretch he received 943 legal questions and answered 939. When I asked him why he hadn't answered the other four, a look of profound exasperation crossed his broad face. "Traffic law," he said. "I'm sorry, I don't know traffic law." By the end of July he was the number 3 rated expert in criminal law on AskMe.com. Beneath him in the rankings were one hundred and twenty-five licensed attorneys and a wild assortment of ex-cops and ex-cons. The next youngest person on the board was thirty-one.

In a few weeks Marcus had created a new identity for himself: legal wizard. School he now viewed not so much as preparation for a future legal career as material for an active one. He investigated a boondoggle taken by the local school board and discovered it had passed off on the taxpayer what to him appeared to be the expenses for a private party. He brought that, and a lot more, up at a public hearing. Why grown-up people with grown-up legal problems took him seriously was the great mystery Marcus didn't much dwell on—except to admit that it had nothing to do with his legal training. He'd had no legal training, formal or informal. On the top of the Arnold family desk was a thin dictionary, plus stacks and stacks of court cases people from AskMe who had come to rely on Marcus's advice had mimeographed and sent to him, for his review. (The clients sent him the paperwork and he wrote motions, which the clients then passed on to licensed attorneys for submission to a court.) But there was nothing on the desk, or in the house, even faintly resembling a book about the law. The only potential sources of legal information were the family computer and the big-screen TV.

"Where do you find books about the law?" I asked.

"I don't," he said, tap-tap-tapping away on his keyboard. "Books are boring. I don't like reading."

"So you go on legal web sites?"

"No."

"Well, when you got one of these questions did you research your answer?

"No, never. I just know it."

"You just know it."

"Exactly."

The distinct whiff of an alternate reality lingered in the air. It was just then that Marcus's mother, Priscilla, came through the front door. She was a big lady, teetering and grunting beneath jumbo-sized sacks of groceries. A long box of donuts jutted out of the top of one.

"Hi, Marcus, what you doing?" she said, gasping for breath.

"Just answering some questions," he said.

"What were you answering?" she asked, with real pleasure. She radiated pride.

"I got one about an appellate bond—how to get one," he said. "Another one about the Supreme Court. A petition to dismiss something."

"We got some chili cheese dogs here."

"That's cool."

Priscilla nipped into the kitchen, where she heaped the donuts onto a plate and tossed the dogs into boiling water. Foodstuffs absolved of the obligation to provide vitamins and minerals cavorted with reckless abandon. Strange new smells wafted out over the computer.

"Where did you acquire your expertise?" I asked.

"Marcus was born with it!" shouted Priscilla. Having no idea how to respond, I ignored her.

"What do you mean?" Marcus asked me. He was genuinely puzzled by my question.

"Where does your information come from?"

"I don't know," he said. "Like, I really just don't know."

"How can you not know where knowledge comes from?" I asked.

"After, like, watching so many TV shows about the law," he said, "it's just like you know everything you need to know." He gave a little mock shiver. "It's scary. I just know these things."

Again Priscilla shouted from the kitchen. "Marcus has got a gift!"

Marcus leaned back in his chair—every inch the young prodigy—pleased that his mother was saving him the trouble of explaining the obvious to a fool. It was possible to discern certain lines in Marcus's character, but the general picture was still out of focus. He had various personas: legal genius, humble Internet helpmate, honest broker, ordinary kid who liked the Web. Now he cut a figure familiar to anyone who has sat near a front row in school—the fidgety, sweet-natured know-it-all. What he knew, exactly, was unclear. On the Web he had come across to many as a font of legal exper-tise. In the flesh he gave a more eclectic performance—which was no doubt one reason he found the Internet as appealing as he did. Like Jonathan Lebed, he was the kind of person high school is designed to suppress; and like Jonathan Lebed, he had refused to accept his assigned sta-tus. When the real world failed to diagnose his talents, he went looking for a second opinion. The Internet offered him as many opinions as he needed to find one that he liked. It created the opportunity for new sorts of self-per-ceptions, which then took on a reality all their own.

There was something else familiar about the game Mar-cus was playing, but it took me a while to put my finger on it. He was using the Internet the way adults often use their pasts. The passage of time allows older people to remember who they were as they would like to have been. Young peo-ple do not enjoy access to that particular escape route from their selves—their pasts are still unpleasantly present—and so they tend to turn the other way and imagine themselves into some future adult world. The sentiment that powers their fantasies goes by different names—hope, ambition, idealism—but at bottom it is nostalgia. Nostalgia for the future. These days nostalgia for the future is a lot more fash-ionable than the traditional kind. And the Internet has made it possible to act on the fantasy in whole new ways.

Priscilla shouted from the kitchen: "Marcus had his gift *in the womb. I could feel it.*"

Now Marcus had his big grin on. "Welcome to my brain," he said.

"What?"

"Welcome to my brain."

He'd said it so much like a genial host offering his guest the comfortable chair that I had to stop myself from saying "Thanks." Behind him was a long picture window overlooking the California desert—the view was the reason Priscilla loved her house. Beyond that, brown mountains. In the middle distance between white desert and brown mountain, a parachute ripped open and a body jerked skyward.

"Let's try this again," I said.

"Okay," he said, cheerfully.

"Basically, you picked up what you know from watching Court TV shows," I said.

"Basically," he said.

"And from these web sites that you browse."

"Basically."

Priscilla shouted out from the kitchen, "How many dogs you want, Marcus?"

"Two and some donuts," hollered Marcus.

"What do you think these people would have done if you weren't there to answer their questions?" I asked.

"They would have paid an attorney," he said. But as he said it his big grin vanished and a cloud shadowed his broad, open face. All of sudden he was the soul of prudence. It may well have been that he was recalling the public relations fiasco that followed the discovery by a hundred or so licensed attorneys on AskMe.com the true identity of the new expert moving up their ranks. In any case, he lifted his giant palms toward me in the manner of the Virgin Mary resisting the entreaties of the Holy Spirit, and said, "Look, I'm not out there to take business away from other people. That's not my job."

"But you think that legal expertise is overrated?"

"Completely."

Once Marcus attained his high rankings on AskMe.com, a lot of people he didn't really know began to ask for his phone number and his fee structure. For the first time, for some reason he was unable to explain fully, his conscience began to trouble him. He decided it was time to come clean with his age. To do this he changed his expert profile. Where it had read "legal expert," it now read "15 year old intern attorney expert." A few hours after he posted his confession, hostile messages came hurtling toward him. A few of them came from his "clients," but most came from the lawyers and others who competed with him for rankings and prizes and publicity. A small war broke out on the message boards, with Marcus accusing the lawyers of ganging up on him to undermine his number 3 ranking and the lawyers accusing Marcus of not knowing what he was talking about. The lawyers began to pull up Marcus's old answers and bestow on them lowly one-star ratings—thus dragging down his average. (At the time, third parties could score expert answers; after the incident AskMe.com changed its policy.) Then they did something even worse: asked him detailed questions about the finer points of the law. When he couldn't supply similarly detailed answers, they laid into him. Marcus's replies to the e-mail lashings read less like the work of a defense attorney than of a man trying to talk his torturers into untying him:

"I am reporting your abusive response, for it hurts my reputation and my dignity as an expert on this board."

"Please don't e-mail me threats."

"You really are picking on me."

"Leave me alone! I am not even a practicing attorney!"

"Please, I beg of you, stop sending me letters saying that you'll be watching me, because you are scaring my parents."

"I really just want to be friends."

"Can't we just be friends?"

To which Marcus's wittiest assailant replied, "In your last two posts you've ended by asking that I be your friend. That's like the mortally wounded gladiator asking to be friends with the lion."

On the one hand, the whole episode was absurd—Marcus Arnold was a threat to no one but himself and, perhaps, the people who sought his advice. To practice law you still needed a license, and no fifteen-year-old boy was going to be granted one. At the same time, Marcus had wandered into an arena alive with combustible particles. The Internet had arrived at an embarrassing moment for the law. The knowledge gap between lawyers and non-lawyers had been shrinking for some time, and the Internet was closing it further. Legal advice was being supplied over the Internet, often for free—and it wasn't just lawyers doing the supplying. Students, cops, dicks, even ex-cons went onto message boards to help people with their questions and cases. At the bottom of this phenomenon was a corrosively democratic attitude toward legal knowledge, which the legal profession now simply took for granted. "If you think about the law," the past chairman of the American Bar Association, Richard S. Granat, told the *New York Times*, in an attempt to explain the boom in do-it-yourself Internet legal services, "a large component is just information. Information itself can go a long way to help solve legal problems."

In that simple sentence you could hear whatever was left of the old professional mystique evaporating. The status of lawyering was in flux, had been for some time. An anthology that will cause elitists to weep will one day be culled from the long shelf of diatribes about the descent into mass culture of the American lawyer at the end of the twentieth century. Separate chapters will detail the advent of the billable hour, the 1977 Supreme Court decision permitting lawyers to advertise their services, and a magazine called the *American Lawyer*, which, in the early 1980s, began to publish estimates of lawyers' incomes. Once the law became a business it was on its way to becoming a commodity. Reduce the law to the sum of its information and, by implication, anyone can supply it. That idea had already traveled a long way, and the Internet was helping it to travel faster. After all, what did it say about the law that even a fifteen-year-old boy who had never read a law book could pass to a huge audience as an expert in it? It said that a lot of people felt that legal knowledge was accessible to the amateur. Who knows: maybe they were right. Perhaps legal expertise was overrated. *Completely.*

By its nature the Internet undermined anyone whose status depended on a privileged access to information. But you couldn't fairly blame the Internet for Marcus Arnold, any more than you could blame the Internet for Jonathan Lebed. The Internet was merely using Marcus to tell us something about ourselves: we doubted the value of formal training. A little knowledge has always been a dangerous thing. Now it was becoming a respectable thing. A general collapse in the importance of formal training was a symptom of post-Internet life; knowledge, like the clothing that went with it, was being informalized. Casual thought went well with casual dress.

Technology had put afterburners on the egalitarian notion that anyone-can-do-anything, by enabling pretty much anyone to try anything—especially in fields in which "expertise" had always been a dubious proposition. Amateur book critics published their reviews on Amazon; amateur filmmakers posted their works directly onto the Internet; amateur journalists scooped the world's most powerful newspapers. There was no reason licensed professionals shouldn't be similarly exposed—after all, they were in it for the money just like everyone else. In late 1999 an outfit in Boston called Forrester Research got the novel idea of going out in the field and talking to doctors about the Internet. The subsequent report was called "Why Doctors Hate the Net." There were many reasons. Patients were pestering doctors to offer advice by e-mail so they didn't have to spend hours hanging around waiting rooms. The doctors hadn't figured out how to get paid to answer e-mails. But if they failed to answer e-mail the patients turned elsewhere for their answers. That was another reason doctors hated the Net: the patients were getting uppity. The Net encouraged patients to believe they knew more than they did. They'd walk into the doctor's office in a state of high indignation armed with some printout from an Internet health care site that utterly disproved that the pain in their gut was, as the doctor had claimed, gallstones. The doctors took their mystique even more seriously than the lawyers—in part because that mystique actually helped them to cure people—and they were displeased to be challenged. But at least their business was not directly threatened. No one was going to put his life in the hands of some web site. Not yet.

And so the situation in which Marcus Arnold found himself in the late summer of 2000, while bizarre, was revealing. Marcus had been publicly humiliated by the real lawyers, but it didn't stop him from offering more advice. He clung by his big mitts to a lower ranking. Then the clients began to speak. With pretty much one voice they said: Leave the kid alone! A lot of people seemed to believe that any fifteen-year-old who had risen so high in the ranks of AskMe.com legal experts must be some kind of wizard. They began to seek him out more than ever before; they wanted his, and only his, advice. Marcus wiped himself off and gave it to them. In days his confidence was fully restored. "You always have your critics," he said. "I mean, with the real lawyers, it's a pride issue. They can't let someone who could be their son beat them. Plus they have a lot more time than I do. I'm always stretched for time. Six hours a day of school, four hours of homework, sometimes I can't get on line to answer the questions until after dinner."

In spite of this and other handicaps, Marcus's ranking rebounded. Two weeks after he disclosed his age, he was on the rise; two weeks later he hit number 1. The legal advice he gave to a thousand or so people along the way might not have withstood the scrutiny of the finest legal minds. Some of it was the sort of stuff you could glean directly from Judge Judy; more of it was a simple restating of the obvious in a friendly tone. Marcus didn't have much truck with the details; he didn't handle complexity terribly well. But that was the whole the point of him—he didn't need to. A lot of what a real lawyer did was hand out simple information in a way that made the client feel served, and this Marcus did well. He may have had only the vaguest idea of what he was talking about and a bizarre way of putting what he did know. But out there in the void, they loved him.

Marcus's father, Melvin, worked at a furniture retail outlet two hours' drive from home, and so wasn't usually around when his son was handing out advice on the Internet. Not that it mattered; he wouldn't have known what Marcus was up to in any case. "I'm not the sort of person who gets on the computer," Melvin said when he arrived home and saw Marcus bashing away for my benefit. "I *never* get on the computer, as a matter of fact." And he said this matter-of-factly, in a spirit in no way defiant or angry, just gently resigned to the Way Things Are. "When I need something from the computer," he also said, "I ask Marcus."

"It just gives me more computer time," said Marcus, and resumed his furious typing.

What with the computer smack in the center of the place, the Arnolds' house didn't allow me to talk to Melvin without disrupting Marcus. When Marcus realized that he was about to be forced to listen to whatever his father might have to say about his Internet self, he lost interest. He called for Marc, and the twin bear-boys lumbered out the front door. On the way out he turned and asked me if I knew anyone in Hollywood he might talk to. "I think what I really want to do," he said, "is be an actor." With that final non sequitur, he left me to cross-examine his parents.

The first thing that was instantly clear was that, unlike

their son, they were aware that their lives were no longer what anyone would call normal. The Lebeds had proved that if your adolescent child was on line, you didn't need to leave your house to feel uprooted. The Arnolds were already uprooted, so they didn't prove anything. They'd moved from Belize to South-Central Los Angeles. They'd moved from there to Perris for a reason, which Melvin now calmly explained to me. At the family's Los Angeles home Marcus's older brother had been murdered. He'd been shot dead in cold blood by an acquaintance in the middle of a family barbecue. The man who shot him had avoided the death penalty. He was up for parole in 2013. "Marcus didn't tell you about that, did he?" asked Melvin, rhetorically. "In my opinion that's how Marcus got interested in the law. He saw that it wasn't fair."

The Arnolds had moved to Perris shortly after their son's murder. Not long after they'd arrived, Marcus asked for a computer. He'd waited until he crashed the top ten on AskMe.com before he let his parents know why, suddenly, he was up all hours bashing away on the family keyboard. His parents had had radically different reactions to the news. His mother nearly burst with pride—she always knew that Marcus was special and the Internet was giving him a chance to prove it. His father was mildly skeptical. He couldn't understand how a fifteen-year-old boy could be functioning as a lawyer. The truth is, Melvin hadn't taken Marcus all that seriously, at least not at first. He assumed he was reacting to the grief of the murder of his older brother. Then the phone started to ring . . . and ring. "These were grown-up people," said Melvin, still incredulous at the events taking place under his roof. "They call this house and ask for Marcus. These people are like forty, forty-five years old and they're talking to Marcus about their legal problems, but they're not including the parents. That's where I get scared, because it's not supposed to work like that."

"Well . . ." says Priscilla. She scrunched up her big friendly face in what was clearly intended to be disapproval. "They're not acknowledging the fact that he's fifteen. They're acknowledging the fact that he can give them some legal advice."

"But the phone," Melvin said, "it is always ringing. These people want Marcus to give them legal advice. I mean, really, it's like what he does, people do as a *job*. And he's doing it right here. I get so frustrated. I always say, 'Marcus, you're talking too much, you're talking too much.'"

"But that's what attorneys do," said Priscilla, "they talk a lot."

Melvin gave up on his wife and turned to me to explain. "I tell him to stay off the phone, stay off the computer. This is the thing I keep on saying to him. Nobody else in this house can ever use the phone. There's no way I can stop him, but still . . ."

"But attorneys talk—that's what they do," said Priscilla.

"I don't use the phone anyway, really," said Melvin. "The calls come, they're never mine, you know. It's always Marcus, Marcus, Marcus—people calling him from everywhere."

They were off and running on what was clearly a familiar conversational steeplechase. "I don't understand," I said. "How do all these people have your phone number?" But neither of them was listening. Priscilla, having seized on her main point, was now intent on spearing Melvin on the end of it. "But that's what he's got to do," she said. "That's what attorneys do! Talk!"

"Yeah, but he's *not* an attorney," said Melvin. He turned to me again in a bid for arbitration. "He drives you nuts with his talk. Nuts!"

"How do they get your phone number?" I asked again.

"But he will be one day," said Priscilla. "He has that gift."

"He's a kid," said Melvin.

"How did they get your phone number?" I asked, for the third time.

Priscilla looked up. "Marcus puts it on the Internet," she said. To her it was the most normal of things.

Melvin took a different view. Maybe it was the distinct feeling he had that a lot of Marcus's "clients" had had to stand in line at a pay phone to make their calls. Or that they always seemed to prefer to wait on hold rather than call back later. Or their frantic tones of voice. Whatever the reason, he didn't like it. "I told Marcus," he said, wearily, "that we don't even know who these people are, they might be criminals out there, that you're not supposed to give them our phone number, our address."

Priscilla furrowed her brow and attempted to conjure concern. "What really scared me one time," she said, less with fear than in the spirit of cooperation, "was this lady that he was assisting with her criminal case. The lady sent him the whole book of her court case. I said, 'Marcus, why would you want to take this upon yourself, you've got to tell this lady you're just fifteen years old.' But he didn't listen to me. The point came that the lady actually wanted him to go to court with her, and I said, 'No, we've got to stop it here, because you don't have a license for that, you don't study law.' He said, 'Mom, you've got to drive me to the court. I know what I'm doing.' I said, 'No way, you don't have a license to dictate the law.'"

I could see that her heart wasn't in this soliloquy. She stopped and brightened, as if to say she'd done her best to meet her husband halfway, then said, "But I think all of this Internet is good for Marcus."

"Do you think Marcus knows what he's doing?" I asked.

"Oh yes, very much," she said. "Because there's a lot of times that we would watch these court shows and he would come up with the same suggestions and the same answers like the attorneys would do."

That appeared to settle the matter; even Melvin could not disagree. Marcus knew his Court TV.

"Can you see him charging for this advice?" I asked.

"At what age?" Melvin said. A new alarm entered his voice.

"Thirty."

"I hope," said Melvin, with extreme caution. "I hope he will do well."

"He's supposed to have his own law firm by then," said Priscilla.

By the time I arrived at the Arnolds' house I had spent a year wandering around and paying calls on people who had distinguished themselves one way or another on the Internet. You didn't have to look very hard to see that the Internet was playing host to a lot of little social experiments. The trouble with the Internet, from the point of view of a real-world traveler seeking to minimize his discomfort, is that it seized the imagination in the least desirable places. The people who wound up using the Internet to violate some social norm invariably lived in purgatory. Desert towns, strip-mall villages, Nordic wastelands, third-world ghettos—these were the landscapes that inspired a person to seek an outlet from his predicament. There had been this twenty-three-year-old Russian named Andrei Filonov, who lived in a squalid old Soviet-era apartment building on the outskirts of Tallin, which, it emerged, was the capital of Estonia. One hundred thousand Russians who had been essentially left behind when the tanks withdrew had been stuffed into a jungle of decrepit high-rises. They weren't happy about it. When Andrei walked out of his one-room prehistoric flat, he entered a dank concrete hall reeking of urine, thick with anti-American graffiti. Yet somehow, during the spring of 2000, Andrei had figured out that he could advertise his existence on a web site run by a U.S. company called E-Lance. He was instantly hired to write software by rich people and companies in America and Western Europe. Within three months he'd become Estonia's highest-paid programmer. He was thinking he might move to America and start his own software company.

Another young man laundered by the Internet was a cartoonist in Los Angeles named Tom Winkler. Winkler's squalor was self-induced; a visit to his apartment was enough to alter anyone's definition of that pungent noun. Soiled underwear on the floor and empty beer cans on the tables meant that Winkler had just finished his annual cleaning. He had worked as an animator on the television show *The Simpsons*. The job bored him, but no employer had any interest in paying for what he actually wanted to draw. So Winkler went home one day and created his dream job on the Internet: a scatological comic strip called *Doodie*. Each day he drew another cartoon of, as he put it, "the money shot of potty humor." After six months he had an audience of half a million people and offers from film studios to turn his running sketch, which was nothing more than weird shit jokes, into a movie. He'd had to hire a lawyer to handle the negotiations. "People have been deprived of seeing quality potty humor in animation since the onset of filmmaking," Winkler said, "because animation requires a lot of money, and to get the money you need to show your project to some people who will give you the money. And if you show them a big doodie coming out of somebody's butt, they're going to say 'No thank you.'" But if you show them a big doodie coming out of someone's butt to an audience of half a million devoted fans, they say, 'Yes, please.'" All by himself he'd demonstrated that, even in Hollywood, there are still social norms that people prefer not to violate unless you can prove to them that it pays. The Internet allowed for no end of proof-of-concept.

This trip of mine was overlong and underscientific but it had its purpose: to understand what had occurred that would not have occurred without the Internet. Not because I was under any illusion that the Internet had, in a simple sense, caused anything. That would be to mistake a necessary for a sufficient condition. The events I investigated had occurred because the Internet filled some kind of social hole. If the hole didn't exist in the first place, the filling up of it never would have happened.

The biggest and most obvious hole had been what appeared to be a social need to alter former relations between insiders and outsiders. Democracy always craved more of itself. It was as if the world had been hungering for a more extreme individualism and an assault on the organizations that hindered it. The Internet had facilitated this change by enabling the outsiders to ignore a lot of old rules—social mores, market restrictions, in some cases actual laws. It had allowed a lot of people who otherwise never would have been let anywhere near the inside to get close enough to make trouble. Jonathan Lebed and Marcus Arnold were two perverse cases in point. Financial and legal professionals might have been appalled by them, but they were unable to stop them from mocking their establishments. The ease with which the outside tormented the inside had changed the meaning of "outside" and "inside." Outsiders were now encouraged to shape up and organize themselves and become, in general, more self-conscious.

The Internet was, among other things, a wonderful prism through which you could observe the new conversation between outsiders and insiders. By the time it went boom in the fall of 1994 there was already a well-established dialogue between Internet outside and Internet inside—an argument between the corporations that sought to profit from the Internet and the loose groups of individuals vaguely hostile to those corporations, who wanted to use the Internet for antimaterialist purposes. Five and a half years later the dispute came to an interesting boil. On March 14, 2000, a twenty-year-old employee of America Online named Justin Frankel posted on the Internet a piece of software designed to undermine America Online—and a lot of other huge and profitable corporations.

Frankel had made his name when he was a student in Sedona, Arizona, where he grew up, by writing a piece of software that enabled computers to play digital music files. In eighteen months the program, called Winamp, was downloaded fifteen million times. Frankel had built a company around Winamp, called Nullsoft. Everything about Nullsoft except the quality of the software was laced with irony. (The name was intended to insult Microsoft.) Frankel didn't charge for his software; instead he asked those who could to leave a tip behind. A lot of people did, and at the age of nineteen, Justin Frankel found himself on the receiving end of millions of dollars. He hired his father to keep the books. Two years later he sold Nullsoft to America Online for an amount variously reported as $70 million, $82 million, and $100 million, but not before he'd said, "Nullsoft is and was about all these good things that ultimately don't matter to most businesses." The deal left him an employee of AOL.

The new code Frankel posted on the Internet in March 2000 demonstrated that working at AOL hadn't slaked his thirst for uncommercial behavior, or, at any rate, unconventional commerce. Frankel had named this code Gnutella—another convoluted inside geek joke involving the chocolate spread Nutella and a free computer operating system called GNU. Gnutella enabled people to share files, computing power, and anything else that resided inside PCs over the Internet without the help of a central computer server. The trick was putting together someone who had something to share with someone who wanted it. To do this Gnutella employed the logic of a person seeking directions in a crowded room. The Gnutella user sent a query onto the network—for instance, "Do you have a text file of Huckleberry Finn?" The query went first to the nearest other Gnutella user at hand. That user's computer checked its files to see if it had *Huckleberry Finn*. If it did, it passed the book along; if it didn't, it passed the query onto the nearest seven other Gnutella users, who, if they were unable to satisfy the demand for *Huckleberry Finn*, in turn passed it on. The query spread geometrically, and within moments the entire network of Gnutella users could be scoured without ever requiring the assistance of the old Internet authorities, the central servers and companies that controlled them.

That was the critical status detail: the two-tiered class system on the Internet, as it is currently configured. All questions come from the client—the PCs on the fringe of the network—and all answers come from the servers—the big computers at the center of the network. The need for big computers translates into the need for big companies, to buy and control the computers. Frankel's program eliminated the need. Gnutella turned every computer on which it was installed into both a server and a client—"servent" was the word that got dreamed up for what happened to your PC after you downloaded Gnutella. It enabled your personal computer to answer queries as well as make them. Hence the general name for what Gnutella enabled: peer-to-peer computing. Gnutella eliminated the class system and made all computers, if not all computer users, equal. Peer-to-peer computing made the Internet what it wanted to be, and what it was originally designed to be: an exchange of equals or, at any rate, an exchange among people on the same level. A pancake.

Frankel later claimed that he'd created his software to make it easier to find recipes, but his friends knew that that was just Justin being Justin. He'd written the program in the heat of the lawsuit filed by the Recording Industry Association of America (RIAA) against the file-sharing company Napster. Napster had been the first wave of software designed to enable people to swap music—that is, get it for free over the Internet. It had been born in the first half of 1999 inside a college dormitory room in Boston, Massachusetts. By the end of that year tens of millions of people, and an overwhelming majority of university students, used the service. On December 7, 1999, the RIAA had filed a lawsuit against the company for copyright infringement. A few months later a U.S. district court in San Francisco had ruled in its favor. The main effect of this was to infuriate and incite people like Justin Frankel, who thought music and every other piece of intellectual property should be free. That was the most obvious immediate purpose of Gnutella: to share music, and anything else that could be digitized, without using a central server, as Napster did. Without a central server, and a corporation running the central server, there would be no one for the music industry to sue.

Frankel's code was the commercial equivalent of a fart joke at a formal dinner party. It took less than a day for his employer, now at the center of the music industry, to discover what he'd done. When Gnutella went up on the Web, AOL was busy merging with Time Warner. Time Warner owned Warner Music, which in turn controlled a gaggle of music labels. The head of Time Warner, Gerald Levin, called the head of AOL, Steve Case, and asked him to tell his employee to retract his software. In turn Case called Frankel, and Frankel, who now owned a lot of stock options in AOL, complied with the request. Time Warner issued a statement that called Gnutella an "unauthorized freelance project." On Frankel's behalf, AOL systematically declined requests for interviews. The company, which was once on the outside, was now inside. It fashioned itself as part of the wide-open fun-loving New Economy, but in its attitude to information about itself it might as well have been the Pentagon. Of course that didn't stop reporters from trying to pry into the psyche of Justin Frankel. But they had to do this without the assistance of the man himself. The *Washington Post* perhaps drew closest when it tracked down Justin Frankel's mother, in Sedona, Arizona. "My son is really a rebel," she said. "He thinks everything should be free."

The new tighter relations between outsiders and insiders could not have been more eloquently expressed. The renegade employee who behaves as if he is worth $70 million, and in some cases actually *is* worth $70 million, does whatever the hell he wants to do. His corporation tolerates him because it knows that the alternative is worse. It is far better to keep the enemy close, by bribing him with stock options, than to have him out in the wild, foraging. The employee, for his part, is still ever so slightly compromised. He must pay lip service to the boss's demands and agree to take down his software, which makes the boss feel that he is indeed the Boss. This has no real effect other than to keep up the appearance that the chain of command holds firm—but keeping up appearances becomes terribly important when the underlying reality is so precarious. As Justin Frankel well knew, it didn't matter whether his Gnutella software remained on the Web or not. The moment he posted it, word spread on the IRC hacker channels, the Internet chat rooms of choice for people like Justin Frankel. "Nullsoft just released an open-source Napster clone," read a message just a few minutes after the fact. "It does MP3s [music files], movies, and any other format you could want." During the nineteen hours the program was up, ten *thousand* people downloaded it.

Then, just in case the world hadn't figured out that his relations with central authority remained a work in progress, Justin Frankel wrote another piece of software and posted it on the Web. This one deleted the advertisements from AOL's instant messenger service. After that he went back underground, declining all requests for interviews. When you typed in the address for the web site he'd created to spread his software—Gnutella.com—all you got was this single cryptic line:

It's All Okay

Inside of forty-eight hours Gnutella became the rallying point for the Internet outsiders. Justin Frankel hadn't published his source code, which meant that his code couldn't be edited and improved upon by other programmers. But it only took two days for one of the ten thousand people who had downloaded the program, a bright young man in Florida with time on his hands named Bryan Mayland, to back it out. "I was on IRC at the time," Mayland wrote to me to explain how he had done this so quickly, "and people were asking questions about [Frankel's program] so I started answering them. . . . There was one missing piece of information that I couldn't figure out. Luckily, a friendly Gnutella insider appeared briefly on IRC to answer a couple of questions, solving the mystery. . . . Next thing I know I've got the whole thing decoded, I've got the authoritative web site, and our vastly underpowered web server is getting 25,000 hits a day."

Gnutella-the-computer-program wasn't the issue. Oddly, the code had been written with a relatively small number of users in mind. When a judge ordered Napster to shut down four months later, Gnutella was swamped with millions of requests from college students looking for an alternative source for music, and the system failed. "The Napster Flood," the day was now called on the Gnutella network. To a trained computer scientist, Frankel's thought process looked perfectly moronic: you don't build smallness into a system that depends on large numbers of users. And Gnutella, to be really effective, depended on large numbers of users. The more users, the more stuff they'd have to share.

Gnutella-the-idea was the issue. The professorial class in computer science might have hated it, but the hacker class loved it. Ate it up. Justin's concept was brilliant; other, more careful people could sort out the details later. The concept was, among other things, an idea of how brainpower, human and computer, might be harnessed—not by yoking it together inside a big company but by coupling it loosely on the outside, free from corporate control. Later, many people would claim to have been thinking about this notion. But it was Frankel's code that acted as a catalyst for action. Within a few months of the Gnutella release pretty much every major computer company had people researching the newly fashionable field: peer-to-peer computing. There were dozens of start-up companies with an eye on the great prize: reinventing the Internet. Intel and Sun Microsystems had whole groups devoted to the subject. Six months after the fact the chief technology officer at Intel, Patrick Gelsinger, said that the frenzy spawned by Gnutella was "a revolution that will change computing as we know it."

The inside was trying to get its arms around the new outside. But the outside was not something you could ever get your arms fully around. The moment you did, it was no longer the outside. Frankel might have been left gagging on AOL stock options. So what? Within days he had a successor, a twenty-three-year-old Berkeley student named Gene Kan. Kan became the spokesman for the movement, the high priest of the anticorporate underground, until *his* company was bought for $10 million by Sun Microsystems. The outsider's game was obviously in part a financial shakedown of the inside, masquerading as a moral crusade—but there was more to it than that. It was, at the same time, a genuine moral crusade. No matter who sold out there were always newer, younger people waiting to take their place and push the boundaries of the outside farther from the inside. (At least until they were paid huge sums not to.) "Gnutella's not a company," said one of its programmers. "It's a movement."

Together with a small team of web surfers I went looking for an articulate spokesman for the revolutionary point of view. Less than a year after its formation the movement was already too far-flung for one person to investigate it easily. The Napster Flood had fragmented it into what was essentially a lot of small, informal R&D projects. There were dozens of web sites and chat rooms and message boards devoted to Gnutella, and its potential successors. Most of these, while perhaps interesting to a computer programmer, were not terribly interesting to anyone else. The confusion of spontaneity with originality, the self-adoration, the endless inside jokes, the strained attempts to seem *fun:* the Gnutella hacker community wasn't so different in spirit from a lot of new Silicon Valley companies, where the presence of a couple of lava lamps in the lobby and a few beach balls on the shop floor suffices as proof of a rich outer life. They were all Wild Czechoslovakian Swingers. The problem here was the problem of all close-knit status groups—any newcomer will have the same sickening feeling at dinner parties of Washington journalists or Hollywood filmmakers. The members either speak to each other in a language incomprehensible to the outsider, or they address the crowd outside the group in a manner that is embarrassing.

All I wanted was a view of the outside from the inside that didn't cause me to blush on anyone else's behalf. One of my web surfers, David Irvine, thought he had found it: a web site with the unpromising-sounding address, www.Gnutty.co.uk. It stood out from the rest. Along with versions of Gnutella and news of the day concerning Gnutella, it offered a Gnutella-like worldview. It didn't preach to the converted; it was designed to educate the ignorant. A shining intelligence radiated from the thing, and there was a real sense that whoever had built it grasped the broader implications of what he was doing. David went off to track its builder down. A few days later he came back to me.

"I found him."

"Great." He sounded hesitant.

"His name is Daniel Sheldon."

"Great." Something was wrong.

"He says he's committed to what he calls 'the legend that is Gnutella.'"

"Even better. . . . Where is he?"

"Oldham."

"Where the hell is that?"

"Outside Manchester."

"Let's go see him."

"There's a problem," said David. "He's very bright, very articulate."

"So what's the problem?"

"He's got to check with his mother."

Daniel Sheldon was fourteen years old. And his mother, Loretta, was worried. At first she couldn't understand why anyone from outside Oldham, much less anyone from abroad, would be calling her son. She knew Daniel spent a lot of time on the computer, but she had no idea what he did when he was on it. "He hasn't done anything illegal, has he?" she asked, plaintively, before giving directions to her house.

Oldham would be familiar to anyone who has seen any of the countless films set in working-class Britain. The makers of small films have long understood that despair always seems worse when it is British—on top of everything else it's cold and wet. Oldham was a stage set for a dank mood: the uniformly drab redbrick houses, the idle smokestacks, the corner shop windows papered over with aging flyers stuck up in rogue moments of hopefulness, the fences wreathed with barbed wire. Pine Avenue had no pines, Oak Avenue had no oaks. Only tidiness separated Oldham from poverty; but somehow the tidiness only added to the desperate feeling of the place. It stripped the inhabitants of the excuse that they weren't really trying.

Daniel Sheldon lived on a Birch Avenue with no birch trees, in a brick house indistinguishable from the brick houses that ran for miles in every direction. He had no father—

or none that he or his mother cared to mention. His mother, Loretta, was in her late forties. She worked part time as a housekeeper.

We did the usual British thing of tea and biscuits. British tea is not a social drink but a trick for preserving one's privacy in the presence of others. So long as they are drinking tea the British don't mind how little they are actually revealing about themselves to one another. We sat around a glowing orange electric grate, which substituted for a fireplace, and made idle chitchat through a fog of tea steam. Daniel may have been a teenager but he was still, in appearance, a little boy. Cropped dark hair over a wide-open choir-boy face. An impossibly broad and sweet white smile. His favorite books were the *Harry Potter* novels and his favorite movie—the only movie he'd ever *really* liked—was *Willy Wonka and the Chocolate Factory.* "Life hasn't happened to him yet," said his mother, which prompted Daniel to describe how, at school, for one hour each week, he was force-fed what his mother called Life, in the form of a class called Preparation for Adult Life.

"Well, you need that," said Loretta.

"It's absurd," said Daniel. "Last week we had to pretend to be husbands and wives in a spat."

"You should give the teachers a bit of feedback," Loretta said.

"I did," he said. "I told them I didn't need to be patronized."

"Oh Daniel!" she cried. "Not that much feedback."

"Well, it's true."

"You got to use a bit of subtlety. A bit of subtlety to get what you want."

Daniel adjusted the spectacles on the bridge of his nose. He shook his head slowly and smiled. The only thing about him that wasn't a sweet little boy, apart from his mind, were his mannerisms. He had the body language of a wise old man.

"You see, life hasn't happened to him yet," his mother said again. "He doesn't know these things can be useful."

The effects of the tea had waned. Loretta didn't say then—but might as well have—that life *had* happened to her. She had the clenched mouth and narrow eyes of someone who has been fooled by one too many lies and was intent on not making the same mistake ever again. Her senses were conspiring to cut her off from the world. I wanted to know how on earth a poor little boy in Oldham, England, had come to play a role in the Internet underground; she wanted to know why I wanted to know. "We had that letter from Warner Brothers, you see, so of course we're terrified, aren't we?" she said, when I pointed out that she seemed a little suspicious of my motives.

"What letter from Warner Brothers?"

"Oh, I thought you must have known about that."

I turned to Daniel—doing a poor job of seeming terrified—who explained that he had created a web site for Harry Potter. To do this he'd registered the domain name Harry-Potter.org.uk. "The mistake I made was using my real address," he said. "Usually when I register domain names I use a fake address." The real address had enabled the U.S. film company Warner Brothers—owned by AOL Time Warner—to first find Daniel and then threaten to sue him, as it owned the Internet rights to the name "Harry Potter," or thought it did. "The letter just came through the door," said Loretta. "It was my worst nightmare. I didn't want to be sued by Warner Brothers, did I? I just wanted to lead a normal quiet life."

"I remain adamant that it is my right to build a Harry Potter fan site on the Web," said Daniel.

Loretta didn't know about that. But she had a friend at work who knew a solicitor who recommended ignoring the letter, and as Daniel was intent on doing just that, that's what she did. Warner Brothers, perhaps realizing how it might look to others if they were seen jailing children for their enthusiasm for Harry Potter, faded away. Which was lucky, since Loretta couldn't have afforded a lawyer.

Daniel's mother was at once ashamed and defiant about the things she did not have. She picked at her deprivation as she might a favorite old scab. She didn't own a car, she said, and couldn't afford taxis, and so she and Daniel spent hours each day waiting at bus stops. She wasn't able to buy Daniel the education he deserved. A couple of years back, when Daniel was eleven, he had scored well on admissions tests to a fancy local grammar school. But he had the bad luck of applying the year after the Labour Party came to power. British Labour had kept its promise to end the Tory practice of giving extremely bright children whatever money they needed to escape the state school system. The grammar school had offered Daniel a partial scholarship but Loretta had been unable to pay the balance. When you listened to her describe how she had to trundle her gifted eleven-year-old boy from one depressing comprehensive school to the next in search of the least bad, when he was obviously designed for and deserving of the very best, you heard the sound of a heart breaking. You also heard a mother's love.

Daniel sat in the big chair with his feet curled up beneath him and a big smile on his face radiating some kind of intelligence. All the obvious misfortune that surrounded Daniel Sheldon you forgot ten minutes after you met him. I was immediately aware that I was finding it impossible to view him as deprived. It took me a while to realize why.

"Daniel's had no input on the computer from anybody," said Loretta, coming around at last to the first question I'd asked her.

"That's only because there is no one around to have input from," said Daniel. The neighbors, so far as he knew, were computer illiterate. His friends took no interest in the Internet. His school was so far behind on the subject that it was just now asking Daniel if he might build its web site. I must have looked at Loretta because she said, "Don't look at me. I don't know anything about it. I bought the computer, that's all."

That was September 1999. Six months later, in March 2000, Daniel had been fully enough aware of the inner game of hacking that he knew of Justin Frankel's program. He downloaded Gnutella, studied it, and made his web site, all within a few days.

"Why did it ever occur to you to buy him one?"

"I knew I needed to get him one. For educational purposes. I thought I'd let him down if I didn't."

"You had no interest in it yourself?"

"I thought that I might use it from time to time," she said. "But when it came into the house it went into Daniel's room and that was the end of it. For me to go on it I've got to go into his pits of a bedroom, and I'm not going to do that, am I?"

I was becoming an old hand at computer-generated family disputes—both at identifying them and tossing fuel on them, then backing away in the spirit of the arsonist to watch the flames do their work. "So it has caused some tension between the two of you?" I asked.

"Oh, we don't argue about it," she said. "The only problem we ever have is about the telephone line. I can't afford a second phone line. He's on that computer all hours, so my friends can never call me. It's like a desert island around here."

"We used to argue more," said Daniel, helpfully. "You used to tell me I should spend more time socializing after school."

"I just thought you should spend more time outside," Loretta said. I recalled having read that Justin Frankel's mother had told her son the same thing. To which he'd replied, "Outside is, like, overrated." Insiders played outside. Outsiders played inside.

"I socialize at school for six hours a day," said Daniel, "so I don't really need to socialize with my friends after school. I prefer to spend my life focused."

With that Daniel suggested we head off to his place of work. We wound up a tight staircase to a landing with two rooms. His room was a good twenty degrees colder than the rest of the house. His window was wide open and it was yet another miserable day in Oldham. "I usually keep it cold in my room," he said, closing the window, then the door. "It's good for the computer." The machine in question sat on his desk. It was a skeleton. Daniel had stripped it of its plastic sheathing so it could run all the time without overheating. It sat there humming loudly, a lively pile of metal bones on a black desk under a bright spotlight. The room was otherwise completely dark. Daniel curled up in his desk chair with his legs tucked under him and began to tap at the keyboard. A smile of anticipation rose inside him. He was the picture of happiness.

"Does your mother know what you do in here?" I asked.

"Whatever I've told her," he said. "I could tell her I design web sites and I program, but I could be doing anything. I could be hacking into the White House for all she knew." He smiled. "But I'm not."

"Does she want to know?"

He considered this. "Yes. She wants to know what I'm doing with my time, because it is her house and I am her son. But at the same time I don't think she wants to know what I'm doing in case it's anything too bad."

The skeleton on the desk began what was clearly a familiar dance. In about thirty seconds Daniel had pulled up five different IRC chat rooms, where he and everyone else first learned that Justin Frankel had released Gnutella. His favorite site—like his favorite machine—was a no-frills affair. Ugly text streamed across the screen, unadorned by the usual banners and colors one associates with the Web. To get onto IRC you need some minimum level of technical aptitude; it requires special software. "I call this the underground," Daniel said, "but it isn't really. Anyone can come, provided you have the *nous* to get on." A page popped up that qualified that remark. It read:

> This site is private and you are not allowed entry without permission. If you are a member or supporter of a law enforcement agency, software

company, anti-piracy organization, university, or
federal/government investigation teams or you
were a member at one point you are not permitted to access this site and may not access any
of the material on it, regardless of intentions.

"This is where I get books," Daniel said.

Here libraries of pirated books scrolled across the screen
twenty-four hours a day. Here, a month before, he had stumbled upon *1984*, downloaded it, and read it on his computer screen. He had been interested to know where the phrase
"Big Brother" had come from. When he finished *1984* he
had a new interest in Orwell, so he went on line and found
Animal Farm.

One day someone will make a fortune by locating the
region of the fourteen-year-old brain that enables it to read
a novel start to finish without being distracted by the five
conversations it is simultaneously conducting. One side of
Daniel's screen displayed a list of maybe twenty screen
names; the other displayed messages flying by at an incredi-
ble pace. Seldom were these longer than a sentence; more
commonly they were a fragment. The goal appeared to be to
write as ungrammatically as one spoke; the prose was the lit-
erary equivalent of wearing a T-shirt to work. Following even
a single discussion was headache-inducing. Daniel kept—
always kept—at least five of them going at once. The tabs
rowed up at the bottom of his screen and glowed red whenever someone said something that might be relevant to him,
at which point he clicked into the chat room and fired off
his own message.

"These aren't conversations," I said. "These are video games."

"Sort of, except it's real life."

For the next hour or so he flipped back and forth between the chat rooms and the Web, where he searched for free music. He wasn't looking to steal anything; to do that you wanted a high-speed Internet connection, and Daniel couldn't afford one. On Saturdays, the day he pirated music, he took the forty-five minute bus ride into Manchester, to a new cybercafé called Easy Everything. For a pound he was able to use their computers with high-speed Internet connections for up to six hours. At high speeds you could download a song inside of two minutes, instead of the twenty-five it took him at home. He'd stay at Easy Everything for six hours—they threw you off the machines after that—and come home with one hundred and fifty songs. At home during the week he merely cased the joint, compiling lists of songs available on the Internet for stealing next Saturday.

As it happened, a few weeks earlier Napster had been given a long list by the Recording Industry Association of America of music that it must prevent people from stealing if it didn't want to be sued all over again. "Look at this," said Daniel. He went to Napster's site and searched for the name of a popular song, ". . . Baby One More Time" by Britney Spears. The reply came back: "no matching files found." This was just a polite way of saying that Britney Spears's record company had told the RIAA to tell Napster to block access to the song. "Now look at this," said Daniel, and with a touch both of glee and the keyboard, the alternatives appeared. On his computer screen was a long list of web sites: ItalianNapster, DeathNap, WhyNap, OpenNap, and on and on. These were not variations but direct rip-offs of Napster.com. The sites were all over the world—to shut them down would require lawsuits in fifty different countries. Several of them contained more pirated music than Napster itself ever did. The only difference between them and the original Napster is that they were not hamstrung by the need to prevent the piracy of the RIAA's long list of songs. "Other people all over the world just steal Napster's software and put it up on their own servers," Daniel said. "Napster's being Napstered. And Napster can't stop it. If they shut it down one place, it will just pop up in another."

Daniel plugged in his request to one of them and instantly got what he was after. As Britney Spears downloaded onto his hard drive, he went looking to see what books were available for the taking. At the same time, with whatever was left of his brain, he explained how he couldn't imagine having to actually pay for any piece of intellectual property, as it was all available for free someplace on the Internet.

At length he opened a sixth chat room—the one devoted exclusively to the Holy Grail of the Internet outsider, a workable Gnutella. When he did this, as when he opened the other five, his screen name appeared on the right side of the screen: the_dr.

"the-dr. is in the house," someone wrote.

"welcome the-dr.," wrote someone else.

Daniel read down the list of the twenty or so people in the room. He was able to describe their contributions to the Gnutella movement—which mostly involved writing new versions of Justin Frankel's software or publicizing old ones. He had a fair idea of what many of them were working on now. Many of the programmers who remained in the IRC chat room had been there since Daniel first discovered it, a few days after Gnutella was released. Daniel spoke of this great day in grand historical terms, as if it had taken place in, say, the late Middle Ages. His sense of time was not mine. What I considered a small unit he considered a big one. He said months when he meant weeks and weeks when he meant days. Events had occurred in the chat room "a very long time ago" and yet he had only got his hands on a computer eighteen months before. On the other hand, all the other people in the room were years older than he was—most of them in their early twenties—and he thought of them as contemporaries.

"Have you ever met any of these people?"

"No."

"Would you ever want to?"

"Not particularly. I'd rather keep it impersonal."

The purpose of Daniel's web site had been to attract as many followers to the Gnutella movement as possible. The more people who used Gnutella—or one of its successors—the more likely Gnutella would succeed. But Daniel's ambitions now extended beyond mere cheerleading. There'd been two problems with the original Gnutella, and Daniel and everyone else knew what they were. First, it collapsed when too many people were on it at once. A lot of people meant a lot of people with slow Internet connections, and slow Internet connections acted as bottlenecks for the entire network. Slow modems were unable to process the huge number of queries, and so the queries simply came to a halt at their computers. The other big problem with the original

Gnutella was that too few people on the system shared files. Napster required its users to take specific action to avoid doing this. If you downloaded a song through Napster, Napster listed your music files on its server and automatically turned you into a sharer of music. There were millions of Napster users who had no idea that when they were on the Internet their computers were, often as not, being tapped into by others. Gnutella made sharing more voluntary and offered its users no incentives to do so.

Once he'd discovered that the code written by Justin Frankel was unable to deal with millions of users, Daniel mothballed his Gnutty.co.uk web site and created another called Crapinet. Crapinet.co.uk, to be exact. On it he explained to visitors that "we need a transition network to keep us going until something big comes along." The Gnutella movement had fragmented into dozens of little pieces; there were all these little Daniels off working on solutions. Daniel himself had written his own new version of Gnutella. He'd been told by his elders—that is, the twenty-three-year-olds who served as de facto leaders of the movement—that it didn't represent an advance on the status quo. He went right back to work on another. He was very much a part of a system, but that system differed from a hierarchical corporation. It progressed by peer review. There was never any chance that Daniel would receive pay or formal promotion for his work. All he could hope for was peer approval.

And therein lay the magic of the outsiders' approach. The technical problems would be solved; about that there was hardly a shred of doubt. But they would not be solved for profit; they would be solved for the glory of it. Whoever wrote a Gnutella successor that encouraged the billion or so devices connected to the Internet to share their contents communally would be world famous. He would be the man who changed the Internet, forever. He would have taken a technology that was already a powerful force for decentralization and decentralized it a lot more. In the bargain he would probably have put an effective end to intellectual property rights, if not as an idea, then at least as a fact. He would be a legend in the movement. That was why all these smart kids came to this place called IRC and worked for free. That's why they were spending their lives curled up at their desks, without pay: for street cred. Inside the movement— inside the outside—you didn't get street cred by making money. Just the reverse: making money was a sign of rot. Justin Frankel and Gene Kan—they'd both sold out, in Daniel's and a lot of other people's view: Frankel by letting AOL gag him; Kan by making a business of a Gnutella-like search engine and selling it off to Sun Microsystems.

What Daniel loved about Gnutella was the absence, or at any rate the immediate absence, of financial interest. He didn't like what he had seen so far of money's influence, and who could blame him? So far as he could see, money corrupted whatever it touched. Money kept his mother waiting for buses that never came and prevented him from attending schools that could have brought out the best in him. Money put people on different levels, in a way he found unnatural. And money caused people to forget the right reasons for doing a thing. "I'm good at this because I have an interest in the thing," he said, pointing at the web site he had designed that had led me to become interested in him in the first place. "If you're not interested in the thing, and if you're just in it for the money, then you're not really going to have a sense of achievement. That's why real advances aren't made by commercialists. We're out to make a network that isn't ruled by the capitalists. We're out to make a network that benefits us all and isn't governed or monitored or censored by anybody else, just us, and we're in control of the network."

"Something like this has been tried before."

"I'm aware of that."

I do not know whether it was the sight of Daniel getting his books for free, or something in the Oldham drinking water, but suddenly I found myself extolling the virtues of capitalism. I tried to explain to the lad about the importance of property rights, but it was pointless. It wasn't that he had no respect for property rights. He simply could not imagine why they existed. He didn't mind the idea of paying artists for their work. But he couldn't understand why he needed to pay a record company, or a publisher, or a software company, in the bargain. He imagined that one day the middlemen would vanish and the creators of intellectual property might be paid like waiters—with tips directly from the happy customers—which, when you come to think about it, is how they were paid under the old system of aristocratic patronage. That, he as much as said, was my economic destiny: tips.

"Look," he said now, in an unnervingly levelheaded tone, the tone usually reserved for the older person in the discussion. "I'd like to think I was undermining capitalism, but whether I'm actually doing anything to harm the big banks and Coca-Cola and Nike, I very much doubt. I'm not saying overthrow capitalism overnight. Obviously it won't be overthrown overnight, but maybe just bring the pegs a bit equal."

"Then tell me," I said, "what gives you the idea you could

make something work that has already failed in the real world?" But at the same time I found myself thinking: *No need for you to worry about the banks, there's a kid in New Jersey working on them.*

"Well," he said, "look at my own experience. Look how different that is from anything that has ever happened in the real world. I couldn't walk into a traditional business, aged thirteen, and expect a fully paid job, and start ordering people about. But in the digital world I can do that. Have done it. It brings me to the same level."

Socialistic impulses will always linger in the air, because they grow directly out of the human experience of capitalism. The neurotic, high-strung relationship between the outside and the inside was the market's new and improved way of dealing with the problem. Socialism hadn't been killed by capitalism. It had been subsumed by it. The market has found a way not only to permit the people who are most threatening to it their rebellious notions but to capitalize on them. Gnutella was one example of this; the Internet was another. The Internet—or, to be, for once, terminologically precise, the World Wide Web—was one example of an increasingly common phenomenon. It was dreamed up by an academic with an anticommercial streak named Tim Berners-Lee. It was commercialized by a couple of marginal players in Silicon Valley—Jim Clark and Marc Andreessen—at least one of whom (Clark) was a sworn enemy of the big corporation. Five years later it was a mainstream commercial technology, the religion of corporate apparatchiks everywhere, who now walked around spewing the new clichés about the importance of "knowledge sharing," "being webified," and "thinking outside the box." What the insiders did not know but must have sensed was that, from the moment the corporate mind assimilated the Internet, the outside went to work to undermine the Internet. Gnutella was a shot from the outside fired across the bow of the inside. There will be many more.

The Gnutella movement fell under the heading CAPITALISM AFTER THE COLLAPSE OF SERIOUS ALTERNATIVES. So long as capitalism was opposed, there could be no truck with experiments in human liberty such as this one. Now that the system was no longer opposed, it could afford to take risks. Actually, these risks were no luxury. Just as people needed other people to tell them who they were, ideas needed other ideas to tell them what they meant. That's perhaps one reason that people so explicitly hostile to capitalism were given a longer leash than usual: they posed no fundamental risk. People like Justin Frankel and Daniel Sheldon linger on the fringe until they dream up something that has great commercial potential. Then some big company swoops in and buys them, or they give birth to the big company themselves. Inside every alienated hacker who thinks he stands for the "good things that ultimately don't matter to most businesses" there is a tycoon struggling to get out. It's not the system that he hates. His gripe is with the price the system initially offers him to collaborate.

The incentive for the outsider was to attack the inside right up to the moment he was co-opted by it. The incentive for the insider—and this took some getting used to—was to allow yourself to be attacked, and then co-opt your most ferocious attackers, and their best ideas. This is what AOL proved it understood, by first hiring, then tolerating, Justin Frankel. Obviously the atmosphere of constant insurrection was bad for a lot of individual insiders. But the effect on the system as a whole is to make it more stable because everyone winds up working on its behalf, even the people who think they are put on earth to attack it. And in the end, those people give birth to the ideas that in turn give birth to fantastic wealth. The only thing capitalism cannot survive is stability. Stability—true stability—is an absence of progress, and a dearth of new wealth.

Daniel didn't see things this way, of course. He was still in the larval state of outsider rebellion. His loyalties hadn't been tested by a bribe. But as he perched on the chair in front of his computer with an expression of total satisfaction on his innocent face, his future was already being mapped. He will stir up trouble for the next few years. Eventually that trouble will attract the attention of people willing to pay a lot for it. And one day he will be swallowed up by the system he now says he cannot abide in one great big gulp. He will wind up having what by his current standards seems like lots of money. He may or may not become really rich but he will almost certainly be well-to-do. Because he is unusually sensitive and intelligent he will probably behave better than most do once they have money; he will remember where he came from. But he will become part of what he now deplores.

Which does not mean that it pays to ignore what he has to say. In the hands of outsiders, the Internet was being used to tell you something both simple and important: your economic system is not something to be smug about. Just because it won an ideological war does not mean it is perfect, or even good. You might think you knew that already but really you didn't. You've become numb to it just like everyone else. Or almost everyone else. There are still a few people who insist on finding new ways to fight it. They sound as naïve as children; often they *are* children. But that doesn't mean they are wrong, or that they won't, in some sense, win.

Some wise old person once said that in a democracy you never know who the messenger will be. Daniel was saying, in effect, that while you might not be able to put an end to capitalism, you might well be able to infect it with new attitudes. "The whole point of this," he said, "is that it will one day be very easy to share things with strangers. With people who you have never met in your life. You don't even know who they are, there is no way of finding out who they are, who has got your songs, who has got your files, who has got your artwork. That creates just a million possibilities. It could be, say, you're an unsigned artist and you can't get a record deal. You put your music on Gnutella, or some sort of network, and maybe one day a record boss comes and sees your music and signs you up. Or you don't even need a record company. The possibilities are endless."

I left Daniel Sheldon and drove two hours south in the English countryside to have a look at something that had just happened, of which Daniel was unaware, that suggested his theories were more than fantasies. The address I had been given led to a sheep meadow reeking of what smelled like decades of shit. At one end of it was a barn. Inside the barn were four aging rock stars and a handful of computers. Coming through the barn door, I was met by Steve Hogarth, the lead singer of the band Marillion, dressed in the outrageous garb of a rock star—wild red coat and dashing World War I flying ace scarf and assorted jewelry that no human being other than a rock star could wear with a straight face. He couldn't either. He was dressed up as he was only because he was about to appear on television, and it was the duty of the lead singer of a rock band to look like something other than an accountant. If anything, he seemed to be slightly irritated by the obligation. "Oh, hello," he said, when he spotted me. "Come on in. Can I fix you a cup of tea?" And he instantly began to fidget with that embodiment of the middle-class British soul, the hot-water kettle with the electric cord coming out of its side.

Into the room bounded Marillion's keyboard player, Mark Kelly. He had a shaved head and a wild bomb-throwing look in his eye and appeared altogether more rock-starish. But he, too, ruined that impression in an instant. "Did Steve offer you a cup of tea?" he asked, diffidently, in a way that would have made any woman proud to be his mother. In their defense, the place was a gloriously seedy mess, decorated only with a handful of dusty framed gold and platinum records underscored by the words, "From all your friends at EMI." These are the last traces of the ten million records Marillion had sold. EMI was the record company, owned by Time Warner, that ditched them in 1995, when they were officially pronounced washed-up.

It's hard to swallow the fact that a lot of middle-aged rock stars actually grow up into fully socialized human beings. They pay their credit card bills on time, pick up their kids from school, and are able to see the virtues of both a market economy and democratically elected political leaders. The only trace of rebellion or profanity about either of Marillion's two most vocal members was a T-shirt Kelly held in his hand, which he wanted me to have as a gift. It said:

Marillion
Uncool as F*ck

This was a parody of a British clothing store ad and a bowing to the inevitable. If you find yourself in a group of pop music aficionados, all you have to do is say "Marillion" and they'll fall over themselves laughing, as if you had turned up wearing a tie-dyed shirt and bell-bottom jeans. Marillion is one of those bands that is unfashionable even to mention, and the band no longer bothers to pretend this is not so. It had formed back in 1981, enjoyed a brief moment in the sun in the early 1980s, including one huge hit, called "Kayleigh," which inspired tens of thousands of parents in England and Holland and (oddly) Mexico to name their daughters Kayleigh, then faded into obscurity and penury. They had very little to show for their past success. The contracts they had signed with EMI in 1983 left them with essentially no royalties and no copyrights. ("It's sort of a tradition to sign a really terrible record deal when you're young," said Hogarth.) After they were dropped by EMI in 1995, Marillion had signed with a smaller British company, Castle Records, at even less pleasant terms than before. For every CD the band sold—retail price fifteen pounds—it received ten pence (1.5 percent). In the bargain it not only handed over all its copyrights but received no guarantee that the company would market or promote it in any way. In 1997, when the band's second record was ready to be released, the new label informed its members that they couldn't afford to send them on the usual tour of the United States. Suddenly Marillion became aware that it was in a curious position, at least for rock stars. They had no clear market for their services and yet they continued to supply them. It never crossed their minds to do the dignified thing and announce they were breaking up. Instead they were slowly going out of business.

The record industry couldn't see enough potential sales for the next dozen songs to justify handing Marillion the money it needed to create them. The band's music was difficult and downbeat. Hogarth, who wrote his own lyrics, couldn't glance at a tragic newspaper headline without wanting to sing a song about it. To its fans, Marillion's records from the late 1980s and early 1990s were known simply as "the Suicide Albums." Never so big as Genesis, to which Marillion was often compared, and never so cultish as Kiss, with which it once shared a taste for outrageous face paint, the band didn't even have a market for nostalgia about itself. And so Marillion simply continued to create ever less obviously popular music, most of which was all but useless to radio or television. In the end there's only so much you can do with a ten-minute instrumental played by four aging, fattening, depressing, balding Englishmen who can't dance.

Then something happened: Marillion discovered the Internet. The way they tell it, the whole thing was an accident. When he found out that the band could not afford to tour the United States, Mark Kelly went onto the band's fan web site—every band has a fan web site—just to let the people know they wouldn't be coming. A few days later one of the fans, a man named Jeff Woods who lived in Raleigh, North Carolina, wrote Kelly back with an idea: let the fans on the Internet organize and pay for Marillion's tour.

Woods was an otherwise sober middle-aged man with a strange passion for Marillion's music. He'd been to seventy-five Marillion shows, and proposed to his wife at one of them. He knew there were at least a few others as crazed about their music as he was. "I'd see the same core group at every concert," he said. "I think even the band started to recognize us as these odd people who turn up at every show." He also knew that there were 2,000 Marillion fans on an e-mail list designed by a Dutch fan. ("You think I'm crazy about the band, you ought to see the Dutch.") The group e-mail list was a chat room by other means. Whenever any one of the 2,000 had anything to say about Marillion, they sent an e-mail to the other 1,999. Woods sent out an e-mail with a bank account number and a note encouraging the others to help pay the band's travel expenses. "There was nothing in it for the Freaks," says Woods. (Marillion's fans call themselves Freaks. Don't ask.) "The only guarantee was that if we raised $60,000, the band would come to North America; if we didn't raise the money, anyone who had given money would get their money back."

Kelly had told Woods that it was a nice idea for the fans to come together, but he hadn't taken him terribly seriously. Three weeks later Woods e-mailed him to say that he had raised a third of the $60,000 the band needed. Looking up from the teakettle, Steve said, "Mark came to me one day and said there's all these guys in America, they've opened a bank account and they're collecting money, and they've already got twenty grand. I said, 'You're joking.' That was the first I heard of it. I'm the singer in the band, I don't know what was going on."

It took several months but the fans eventually came up with the $60,000. In early 2000 the band played twenty U.S. cities to good crowds, and the whole thing was bought and paid for by its fan club on the Internet, which was rapidly reshaping itself into a fighting force. That gave the band an idea: instead of dealing with record companies, who were invariably harsh and mercenary and unpleasant, deal directly with the fans, who were nice and encouraging and useful. "That was the point we suddenly realized, hey, this is a really important thing," recalled Kelly. "We've got all these really keen fans that love our music, and they can be mobilized in this way. I mean, in this case, it wasn't by us, but even so, that's when we really started to look at the whole thing about collecting e-mails and names and addresses."

The timing of this thought was fortunate, as it coincided with the band's delivery of the final record on its contract with Castle Records. The band found that the record company's indifference to their fortune gave them a bit of leeway to do things their own way. They scratched the original title of the record and named it instead after their newly designed web site: Marillion.com. Inside the new CD's plastic case the band inserted a form that told the buyer that if he mailed the band his e-mail address, the band would mail him a second disc with previously unpublished songs. Eighteen thousand people took them up on the offer. When the dust settled on Marillion.com, the band's database had swelled to 25,000 names. At the same time they had no deal with a record company—and no prospects for a deal. Marillion reckoned that a record would cost them about £100,000 to produce. It e-mailed its fans an offer: advance Marillion the money to make another record, and the fans would get the record early, and a minuscule picture of themselves on the dust jacket. Within four days 5,000 Marillion fans had agreed to pony up sixteen pounds apiece, in advance, for the record. In a few weeks the band had raised over £200,000 from 16,000 fans—the biggest advance it had had in years, with no strings attached.

The next step was to use its new relationship with the fans to cut a new deal with the record industry. Here was how Steve Hogarth remembered that delicious moment:

> So we went to EMI and said, "Hello, remember us? We've got a record here, you know, and it's very good, it's going to be a world-class album, we're very excited about it, and you can have it for . . . free. And all you have to do is sit back and let us manufacture and sell however many copies that are preordered, and we'll have a cut-off point at the end of this year after which people can't order the album from us anymore. We will fulfill those orders on the understanding that when the album comes out, on the day of release, we won't sell any more records to anybody. So at no point will our version of the album be competing with yours and you can basically pick up the remainder of the sales from the day of release." And they said, "Yes, please!" So we said "Yeah, but there's a catch, you know, we want a very good royalty rate," and they went, "Ooh, how much?" and we went, "Well . . ." and

they said, "We'll have to think about it." And we said, "We also want a decent marketing budget written into the deal." In the end they came back to us and said, "Okay, that's great."

That was the first encouraging new sound heard out of Marillion in a decade. It was the sound of the tone changing in a commercial relationship. The music industry, like every industry, is a small world, and dozens of other bands had heard what Marillion had done and called to ask how to replicate the network. To each band that called, Kelly gave the same advice: "Every time you play a show, try and collect some e-mail addresses, try and build that database, because if you can be in touch with your fans it empowers you right from the start, where you might end up never having to sign a record contract." To which Steve would add, "A traditional record deal is something to run away from in horror. If you take the trouble to get to your fans on the Internet, you can basically go to the record companies, and say, 'Hello, this is our music, I'm the artist, this is my thing, do you want to work for me a little while?' rather than the other way around."

After that the barn that had doubled as Marillion's recording studio expanded rapidly, in two directions. On one side the band dragged in an even seedier mobile home and hired a young American named Erik Nielsen to design a web site and continue to organize the fans. On the other they created a warehouse for the band's new Internet-based merchandising business. By formalizing relations with their fans over the Internet, they encouraged, by the by, a booming business in T-shirts and hats and buttons and videos.

"We're already at the stage now," Kelly said, "where just through Internet sales we could be self-sufficient, without a record company at all. The only downside of that from our point of view is that not all our fans are reachable through the Internet, but in the future it might well be the case that they are, and at that point, well, the world is our lobster."

The people who made it all possible—you could call them a community of fans but they were more than that. They were a hive. A teeming, rabid mass of people now invested in their band as directly as fans ever were. You've heard of interest group politics; this was interest group economics. Not long after the band raised the money from the fans to record its record, a disc jockey on Britain's Radio One asked his listeners to e-mail him a song to accompany an upcoming solar eclipse. One of them suggested a Marillion song called "Afraid of Sunlight." "Marillion," the DJ had said, when he saw it, "aren't they dead?" The poor guy had no idea that he was swatting the side of an electronic hornet's nest. He went on to ridicule the band, and anyone who had anything to do with its music, no doubt certain he was playing to his crowd. Inside of twenty minutes his computer crashed. Word of the insult had instantly spread to Marillion's fan web site. For the next three days the DJ's computer system routinely ceased to function under the weight of the outraged e-mails from Marillion fans. Not hundreds, not thousands, but *tens* of thousands of e-mails from people who saw themselves not merely as lovers of the music but as bankers to the band. The DJ eventually caved and invited Kelly onto his show to discuss this strange new phenomenon.

A new pattern of fan support was taking shape. A few months later, on a radio station in Ireland, a fan called in and requested a Marillion song. The DJ ridiculed him and said he knew a good Marillion joke.

> "Knock, knock."
> "Who's there?"
> "Marillion."
> "Marillion who?"

That was the joke. The response was swift, in the usual form of several thousand e-mails. The next morning the DJ apologized and invited Marillion to play on the show.

The shift in power away from record companies on the inside and to activists like Woods and musicians like Marillion on the outside was just the precondition. What was really interesting was what happened *next*. The subversion of the obnoxious established order invited a new tone into the relations between the people who make the music and the people who listen to it. "Paradoxically," says Steve Hogarth, "it's actually much more spiritual than before the Internet was there because you are able to get a mainline cable to people's enthusiasm. That was never possible before for artists. You could do a show and you could feed off the excitement of a show, but it's not quite the same thing as reading what people are saying about what you've done."

The Internet was being tapped by raw emotion, with potentially great economic consequences. All of a sudden it was possible for a rock star to be directly in touch with a group of people he had heretofore regarded as a glorious, screaming, undifferentiated mob. There isn't a rock star on the planet who hasn't come to the front of the stage and shouted something about how he wants to thank the fans because without the fans he wouldn't exist. Then he vanishes into his limousine with the record company executive and doesn't give his fans a second thought. The structure of his business has meant that he's out of touch with the customer. The same is largely true of writers, filmmakers, software designers, and everyone else putatively engaged in cultural or intellectual life. Marillion is maybe the only band that thinks about the fans the way musicians pretend to think about them. From the moment the fans became their de facto bankers and agents, the band discovered, in the presence of its fans, a new feeling in the air. As Hogarth put it:

> There is this new faith. I actually believe that our fans, if they can download something that we're doing from Napster, will feel that they've sort of let us down if they don't pay for it. That might sound a bit airy-fairy and hippie, but it's actually what I personally happen to believe because I've met these people in the thousands and they really care about us, and I think that if they put us out of business, they wouldn't sleep at night. . . . I mean the record company is a bit like someone who bets ten pounds on a horse because they think it might just win and if it does they'll get loads of money back. That's the relationship the record companies have with the artists. The relationship the fans have with the artists, they're a bit like that guy who looks after the horse and

feeds it and trains it and gets it ready for the race. It's a different level of faith. It's about caring rather than just having a bet.

Making the relationship more direct rendered it less rawly commercial. And to the existing order, this was very threatening. The band's view was that if the Internet wound up killing off the record industry, the reason would be that the record industry had become too much of an industry. Artists were never meant to be extensions of big faceless corporations. The technology was just encouraging artists to do what they should have done on their own: shun the safety of the inside and embrace their natural role as outsiders. The center that held their financial interest and divided their loyalties was crumbling.

THREE

THE REVOLT OF

THE MASSES

Jonathan and Marcus and Daniel were very different characters but they had this in common: they'd used the new masks offered up by the Internet to reinvent themselves in a manner that was, from the point of view of central authority, disturbing. They were able to do this only because central authority itself had been thrown into doubt and was ripe to be challenged.

To see the depth of the problem on the inside you only had to watch the attitude of capital, and the grown-ups who used it. Capital is a useful guide to the social observer. The sheer consistency of its behavior—it never does anything but seek the highest return—means that its movement inadvertently tells you a lot about the world. If capital is moving in some new direction, it is because financial incentives, not capital, have changed. What was true of capital was just as true of capitalists. You can detonate a hydrogen bomb in midtown Manhattan and there will still be some guy at Goldman Sachs worrying about whether he should jump to Morgan Stanley for a signing bonus. When you see the brightest lights dropping out of big firms on Wall Street and in the City of London, and going to work for hedge funds and venture capital funds that specialize in putatively high-risk investing—well, it's a sure sign that something is up. What's up is that capital—the old tool of choice of insiders for buttressing the status quo—is being used routinely to undermine the status quo. Capital and capitalists have become edgier, because it pays them to be edgier.

A lot of people believed that venture capital owed its edginess to the distorting influence of the stock market bubble, and that when the bubble burst the venture capital, and the turmoil caused by new technologies funded by venture capital, would subside. That's not exactly what happened. In the stock market's darkest moments world-historic sums still drifted into the hands of venture capitalists. In the 1980s and early 1990s there had been, typically, about $2 billion of venture capital raised each year in the United States. That number had risen to $9 billion in 1995, the year the Internet went boom, and then to $29 billion in 1998. In 2000 venture capitalists in the United States alone raised $103 billion, according to the National Venture Capital Association. Forty percent of it went into Internet-related businesses. Added to that was another $50 billion or so, raised in Western Europe and Asia, up in just a few years, essentially, from zero. The venture capital end of the stock market had collapsed in 2000, and yet 2000 was the best year ever for raising venture capital, in every corner of the developed world. Probably the sums would decrease in 2001, but they would remain fifty times what they had been just a decade ago. The stock market boom and bust only distracted attention from the very real changes occurring beneath it. People who allocated capital for a living believed that something more fundamental than the taste for stock market speculation was driving capitalism forward.

As it happened, I had spent a lot of time with this species. In the fifteen years since I first came to know him intimately he had changed. The smartest capitalists were no longer the ones who did the big deals with established companies and investors on Wall Street. The smartest capitalists were the ones who did the little deals with the companies that threatened the established companies. The old hot-shot capitalist was so narrow-minded you could use his brain to slice salami. All through the 1980s the prototypical hot-shot sat hunched down beneath his desk with his phone pressed to his ear trying to talk some poor schmuck into wiring a billion dollars from one end of the planet to the other so that he could take a little slice out of it. If there was a world outside that telephone, he didn't much care. Not so the new hot-shot capitalist; the new capitalist was an amateur social theorist. The enterprises he funded suddenly had all these cultural implications that he couldn't ignore. The end users of his capital were often planning not just business but what amounted to a tiny social revolution. To understand his role in the world the capitalist often had his own Grand Narrative.

One venture capitalist more than the others had exhibited an almost obsessive need to tell himself a story to justify what he was doing with other people's money. He was a former engineer with Bell Labs and a former technology analyst with the Wall Street firm Morgan Stanley named Andy Kessler. Kessler had spent most of his career wearing a suit to his office at a giant corporation, yet he remained the sort of person who, just for the hell of it, cracked the computer code that governed his satellite dish so that he could see every television station in Australia. He was a strange mixture of insider and outsider. He was forty-two years old and had a wife and four children, but when you sat him down and asked him to explain himself it was like talking to a wildly energetic twelve-year-old who had spent the last four years of his life working out how to cause the most trouble. Kessler had found the answer to that question from 1996 until 2001, when he and a partner, another Wall Street dropout, raised hundreds of millions of dollars from big financial institutions and invested them in small technology companies. They worked from a single room in Palo Alto, California, and made a small fortune.

To do his job, Kessler had found, he needed to become a kind of connoisseur of relations between outsiders and insiders. He'd gone so far as to codify his belief that the big moneymaking opportunities came from cultural change, and that cultural change nearly always got its start as a subtle shift in relations between what he called "the center" and "the fringe."

One day in the summer of 1998, for instance, Kessler had toured a data center—one of those giant warehouses filled with computer servers that currently power the Internet—of a new company called Exodus. There he'd noticed that to shut down a large part of the Internet for a day, set the world record for economic terrorism, and get his picture on the front page of the *Wall Street Journal,* all he or anyone else needed to do was yank a single fiber-optic networking cable out of the side of a single large computer bank. In some ill-defined way, our security was drifting out of the hands of the people at the center we normally look to for our security. Kessler scribbled a single line:

Pentagon General vs. Exodus Data Hosting Center

This cryptic fragment joined other equally cryptic fragments on a list Kessler kept of the conflicts between outsiders and insiders. The list was both a money map and a glimpse of the direction the future wanted to take. By the fall of 2000 it ran for pages, but a few lines served to convey the general idea:

**Madison Avenue Creative Director
vs. E-Mail Spammer
France Telecom vs. Vodaphone
[European cell phone company]
Public Schools vs. Home Schooling
British Broadcasting Corporation vs. BskyB
[Rupert Murdoch's satellite TV company]
Newspaper Publisher vs. eBay Techie
Corporate Security vs. Hacker
Library of Congress vs. Akamai
Car Dealership vs. Carsdirect.com
Microsoft vs. Linux
Washington Post vs. Matt Drudge
Hollywood Distribution System
vs. George Lucas**

Each line on Kessler's list was, in effect, an argument that the world as we knew it was coming undone: something had occurred that had unsettled some old center of power and prestige and empowered the fringe. For example: newspaper publisher versus eBay techie. eBay is the on-line auction house. It enables people to sell everything from thirdhand lawn mowers to Old Master paintings to a rapidly growing national market of bidders. It is a profitable, rapidly growing company. With grim certainty it gnaws away not just at the auction business but at all classified ads. (Why pay for a classified ad when you can advertise to millions of people for free?) Classifieds keep the newspapers in business. The better the eBay technology becomes, the more people gather on eBay's site, and the more eBay undermines the local newspaper business.

Taken as a whole, the list suggested a more general argument: power and prestige and profit were up for grabs. The Internet undermined all the old sources of insider power: control of distribution channels, control of intellectual property, control of information. All of this it already did and it was still slow-moving, and available to less than half the population. Which is to say that all the things the Internet did to undermine central authority today it would do even better and faster tomorrow. "Technology stocks are down," Kessler said, "but technology itself has no clue whether we are in a bull or a bear market, boom or recession. It just marches ahead." Every venture capitalist now understood this—which is why they were newly eager to pony up resources to the outsiders and eliminate the last technical advantage of the insiders: access to capital.

Beneath Kessler's Grand Narrative was a technical metaphor. Look at any network—the phone system, the

Internet—and you see that the little devices on the edge of it have steadily become more intelligent, and the big powerful devices at the center have grown less important. There is an old adage in technology: intelligence always moves to the edge of the network. Twenty years ago the important computer was the big mainframe at the center of some institution; today the important computer is the PC; tomorrow the cell phone or some other handheld instrument even further out on the edge. Five years ago it was assumed that the Internet would run through central servers. Gnutella had made thinkable the notion that the Internet would have no center; it would be a collection of devices on the fringe, talking to each other.

Kessler believed that this tendency in technical networks also existed in human ones. Corporations, for instance. "When you visit big companies," he said, "you see that, increasingly, information is more readily available to the lowly workers out at the edge of the corporation than it is to the CEO. When this happens the command and control structure of corporations copied from successful military campaigns begins to change; the soldier has more information than the generals and is empowered to act on it, rather than having to wait for information to travel up and down the ranks. And the key to all this is the speed of information."

Perhaps to a former engineer it was only natural that he try to reduce social events to lists and charts and graphs; in any case that's what Kessler had done. He'd written a kind of algorithm to describe the relationship between upstarts on the fringe and the incumbents at the center:

1. Rules are established to create order and maintain profits for incumbents. Examples of rules are: social mores, professional licenses, government regulation, locked-up distribution channels.
2. Cheaper technology suddenly allows for the bypassing of the rules.
3. Incumbents are fat and dumb and happy with current monopolistic profits and their general situation, so they bad-mouth any new stuff which threatens their incumbency or profits, or both.
4. Fringe players emerge to use this ever cheaper technology to simply ignore the rules.
5. Fringe companies attract venture capital since there are great profits to be made underselling the incumbents.
6. Incumbents are in denial until their profits are really threatened and/or market share begins to erode meaningfully.
7. Chaos ensues; fringe players are threatened with lawsuits, government regulation, public shaming, etc.
8. Growth at the fringe accelerates, as it is the right way to do business using new technology.
9. Incumbents co-opt the fringe, or fringe players become the new incumbents and seek to establish new rules.
10. Go to 1.

The Internet's gift to the fringe was to expose a lot of commercial and social rules as dated and inefficient. It would expose even more in the future, as it speeded up and became ubiquitous. This relentless pressure from the fringe did not mean that the center would collapse. But to survive it would be forced to remake itself. Its new shape clearly was not good for business, at least not initially. In fact, it was designed mainly for the benefit of a fringe character.

The obvious starting point for the current phase of capitalism was November 9, 1989, with the fall of the Berlin Wall, and the formal collapse of socialism. From that moment there was no need to flavor the free market with a dash of something else. That was the moment it ceased to be controversial to say "greed is good" and became simply assumed that it was. Little pockets of socialism that had been tolerated when socialism posed a threat now, overnight, seemed horribly retrograde. Why have your capitalism diluted when you can have it straight? Since then, as if by some marvelous coincidence, a lot of new technology has arrived to enhance market forces. The Internet is one such technology. It creates new markets and new competition in old markets and helps to put a better price for the consumer on everything—in the extreme cases of intellectual property it was attempting to put a price of zero. In a few short years it pretty much gutted the principles of corporate socialism—jobs for life, employee and customer loyalty, all for one and one for all—and replaced them with something more raw. It did this in the name of efficiency; just as, in the name of efficiency, it had co-opted the socialistic impulses that could be put to use.

On any time line that describes this phase of capitalism, you would have to include (in addition to November 9, 1989) April 4, 1994 (birthday of Netscape), November 10, 1994 (birthday of Amazon.com), May 5, 1996 (birthday of eBay), January 18, 1999 (birthday of Napster), March 14, 2000 (release date of Gnutella). You would also have to include August 4, 1997. August 4, 1997, was the beginning of the end of another socialistic aspect of modern life: the mass market. Forty years from now, when you have your grandson on your knee and he asks you, "Grandma, how did one hundred million people ever let themselves be talked into buying the same brand of toothpaste?" or "Why did people with things to sell ever think it mattered what country they were selling them in?" you'll say, "Well, you have to know how things were before August 4, 1997."

That was the day a pair of Silicon Valley engineers named Jim Barton and Mike Ramsay started their own technology company. They had no idea what that company might do. It didn't matter: the Internet Boom had entered its most self-indulgent phase. All over Silicon Valley engineers were founding companies before they had any idea of what their companies might do; the urge to innovate preceded the actual innovation. It was widely assumed that even ordinarily smart engineers with the desire to create something new could do so with impunity, and Barton and Ramsay were more than ordinarily smart. They were so smart that a pair of venture capital firms—New Enterprise Associates and Institutional Venture Partners—advanced them several million dollars to get them started, few questions asked. "Three million dollars was pocket change," Ramsay explained.

Barton was an American Midwesterner in his early forties and Ramsay was a Scot in his early fifties, but their identities had nothing to do with the wheres and whens and how longs of their lives. Simple demographics were irrelevant to who they were. They were technologists in search of an idea. Their first idea was to turn the modern home into a network. Computer people have long imagined that the ordinary American home one day would be fully networked, leaving everyone else to wonder exactly what that means. Will the refrigerator order fresh milk directly from the grocery store? Will the furnace and the fish feeder and the vacuum cleaner respond to commands from the office desktop? Anything is possible. That was the good part about home networking as a business idea: the Internet had made it feasible. The bad part about the idea was that it was hard to see the point of it. Oh, it was easy enough to get worked up about it with a fellow geek, but Ramsay and Barton discovered they couldn't explain their dream to anyone else. Ramsay put it this way: "When you build a company around a technology and someone says, 'Tell me again what this thing does?' you need to be able to say, 'It does this.' We found that we couldn't say what home networking did."

And so, after a few months, Barton and Ramsay abandoned home networking. They went back to their venture capitalists and told them that home networking was a bad idea because they couldn't explain it to anyone but other geeks. They had another idea, though. Instead of transforming the entire modern home, they decided to focus on the one appliance that was the closest thing to the center of attention in the modern home: the television.

Barton had become obsessed with the television a few years earlier, when he'd worked at what was then the hottest computer company in Silicon Valley, Silicon Graphics. In the early 1990s Time Warner, AT&T, Microsoft, Silicon Graphics, and other big technology and media companies fell in love with the same idea: that they could change the way Americans watched television. A new device—variously known as the telecomputer, interactive television, or the black box—could be plonked down on top of the American television to offer the viewer an entirely new experience, one in which he would be able to e-mail, shop, and access a virtual library of movies from his couch. There ensued a mad scramble, and Barton was a part of it. He helped to build the only interactive television that actually worked, installed in late 1994 by Time Warner in four thousand homes in Orlando, Florida, and then watched in dismay as his beloved project was overrun by the Internet. The Internet did a fraction of what the new TVs promised, but at a fraction of the price.

Of the few people who dwelled on the way the Internet had swamped interactive television, Barton may have dwelled on it the most. Like a lot of really smart engineers, he had the air of a man used to figuring things out quickly. When you asked him a question, a little smile, and just a hint of self-satisfaction, flickered beneath his light brown mustache and reminded you, gently, that he knew a lot more than the answer. But the TV gnawed at him, precisely because he didn't have the answer. He had sunk the better part of three years into building Silicon Graphics' interactive television, and it had been a commercial disaster. The box worked. And yet no one cared. There were several lessons in this:

No. 1: A big company is not necessarily the best place to create a revolutionary technology.

No. 2: Brilliant gadgets for a mass market do not go anywhere if the masses cannot afford them.

No. 3: The whims of the American consumer are the eighth
        wonder of the world. They can play havoc with the
        most powerful establishments.

When Barton and Ramsay returned to the television,
they had in mind another black box, less technically but
more socially ambitious than the interactive television. They
called it a personal television receiver, but never mind about
that. It was a black box. The main thing about the black box
was that it had a memory—the ability to store programs. It
could record any program as it was watched, as well as anything its owner instructed it to record. This is, of course,
what VCRs were designed to do but didn't, since no ordinarily intelligent person, not even a geek, could figure out
how to make them work. The new box would be simple to
program. It was a VCR that did what it was supposed to do,
even if you were a moron. But it was far more versatile than
that. The viewer could record a great many hours of programming. Or he could simply tell the box to go out and
find him the kind of programs he liked. If he liked unsubtle
and indiscreet women, he could record and store every
episode of *Sex and the City*. If he liked intelligent blood and
guts, he didn't need to wait until TNT's Clint Eastwood
Week—he could just instruct his black box to fetch Clint
Eastwood movies as they played. Once the box was up and
running, the viewer's only constraint on choice was that the
program had to be broadcast by someone, sometime.

The black box also enabled the viewer to treat all television—even live television—as television he had recorded for
his own private use. By pressing a button he could skip the

credits, the huddles, the time-outs, the weather, the endless clicking of the *60 Minutes* stopwatch, and all the other boring stretches of television designed by producers to lull the viewers into watching ads. He could also skip the ads.

Over time, the viewer would create, in essence, his own private television channel, stored on a hard drive in the black box, tailored to his interests. His ability to do this would depend on the amount of computer memory in this box. At the start, Barton reckoned, a black box that cost $1,499 would be able to store about twenty-eight hours of programming; one that cost $699 would be good for six hours. But with the price of computer memory falling by half every eighteen months, the price of the box would plummet: in less than a decade, a black box costing no more than $100 would be able to store the equivalent of an entire Blockbuster Video outlet. In the distant future—a decade hence—the technology would be free. It would be built into satellite and cable boxes. The consumer wouldn't even see it.

There was one other cool thing that the black box did—though Barton didn't appreciate how important it was until later. While the viewer watched the television, the box would watch the viewer. It would record the owner's viewing habits in a way that TV viewing habits had never been recorded. The viewer's every decision would be stored in a kind of private museum of whims. Over time, the box would come to know what the viewer liked maybe even better than the viewer himself. It would know when he turned his TV on and off. It would know when he changed channels; it might even know why. It would know how long he was likely to watch any particular show, and whether he preferred the beginning or the end. All by itself, it would go and record shows that it calculated the viewer might like to watch. The box was more than a box, it was a butler, and the more it learned about its master's whims, the more it would be able to fetch what its master wanted.

The box had certain advantages over every other attempt to transform the television—and there had been many. One was its phenomenal simplicity. Unlike, say, the VCR, it required almost no technical aptitude. The black box would turn the television into a computer, but without making any computer-like demands on the viewer: all the consumer would see was a slightly busier remote control. Another advantage was price. A final advantage was that you could explain it all to an ordinary human being. When someone asked Barton or Ramsay, "Tell me again what this gadget does?" they now had a simple answer: "It lets you watch anything you want to watch when you want to watch it."

Ramsay and Barton decided that in spite of appearances, TiVo, which is what they decided to call their new company, was not a maker of black boxes but a service for people who owned black boxes. TiVo would help each and every American to create his own private television channel. Of course, in the beginning, they would need to build the black box and sell it to the masses. But the black box was not where the money was—the box was, in fact, a big money loser. To kickstart the market, Ramsay, now CEO, and Barton, the chief technology officer, would need to pay some consumer electronics company like Sony or Philips to manufacture the black boxes and to sell them below cost. The trick was to get as many black boxes into the American home as possible.

Once the new boxes were proved to delight their audience, TiVo would then offer its services to the masses: the company's programming software would be in millions of new homes either in tandem with existing cable boxes or, in the future, embedded in new TV sets, cable boxes, or satellite receivers made by companies like Sony or Philips. Thus, the long-term goal of the black box was to become unnecessary. "We'll know we've succeeded when the TiVo box vanishes," Barton says.

The ambition of the thing was breathtaking. The company intended to plop itself down between the 102 million U.S. homes with televisions and the $50 billion U.S. TV industry. And its ambition wasn't limited to the United States; right from the start Ramsay and Barton planned to dish out boxes in Western Europe and Asia. Once the box was in place, TiVo would be the hub of the global television industry. The company would come to know the subtle preferences of each and every television viewer. It would then be able to charge a fee to anyone who wanted to locate TV viewers or groups of viewers: networks, cable companies, advertisers. The trick was to get the box into those 102 million U.S. homes—the easiest homes on the planet to sell new gadgets to—and that would cost money. Lots. Ramsay went back to the venture capitalists and told them that he and Barton needed to lose between $300 million and $400 million before they became profitable. Prior to the Internet boom, the capitalists were chary about sinking one-tenth of that sum into a small, risky venture; now they didn't think twice. "Instead of saying 'No,'" says Ramsay, "they said, 'Great.'"

What made the enthusiasm of TiVo's financial backers even more astonishing was that a rival company had already sprung up. A young entrepreneur named Anthony Wood had stumbled on the same idea as Barton and Ramsay at roughly the same time. Wood, who made a lot of money in computer games, had been frustrated by his inability to persuade his VCR to record episodes of his favorite show, *Star Trek*. He saw the same big trends that lit a fire under Barton and Ramsay: the falling price of computer storage, the TV viewer's desire for choice, the continued inability of Americans to program their VCRs. In early 1998, not long after Barton and Ramsay got their first financing, Wood generously agreed to accept $8 million from the venture capitalists Kleiner Perkins Caufield & Byers and Paul Allen's Vulcan Ventures. He called his new company Replay Networks.

Which is to say that another mad scramble to transform the television was under way, but this time it was more attuned to the spirit of the marketplace—the approach came from the fringe rather than the center. "This is the Trojan horse for the computer industry to gain control of the entertainment industry," said Marc Andreessen, a Netscape cofounder who invested his own money in Replay. "It is the first box built by Silicon Valley that is compelling enough that people want to hook it up to their TV sets." And it was. The moment you had one you couldn't understand why you hadn't had it all along.

The new companies were proposing to do politely to the television industry what Napster was about to do to the music industry: help consumers to help themselves to entertainment without "paying" the networks and advertisers, by watching the ads. Naturally, this disturbed the television net-works and advertisers. In early 1999 Stacy Jolna, TiVo's liai-son with the networks, appeared on a panel before the National Association of Broadcasters. Jon Mandel, an ad executive with MediaCom, was also on the panel. "He start-ed by calling me and everyone involved with this technology 'the devil incarnate,'" Jolna says. "And he went on from there. The basic attitude of the U.S. TV executives was that we were somehow going to destroy a $50 billion business model."

In March 1999 the first TiVo and Replay boxes went into U.S. electronics stores. A Replay box with thirty hours of storage cost $499. A TiVo box with thirty hours of storage cost $399—but then the company generally charges a sub-scription fee of $9.95 a month. By June 2000 the companies had sold about 100,000 boxes between them, and they had done so largely without advertising their products. By June 2001 they had sold 300,000, and the box was experiencing one of the fastest uptakes ever of a new piece of consumer electronics. Microsoft announced it was entering the busi-ness, with an identical product it called Ultimate TV. Sever-al market analysts estimated that TiVo and Replay—or some yet-to-be-created competitor—will have sold five to seven million boxes by the end of 2002—and that within a decade they would be in 90 million U.S. homes. But that was just guessing, and probably a great exaggeration of the truth. No one knows how quickly the companies might arm the entire American population, or even if they would do so.

That was the thing about this technology: it didn't need the companies to succeed. Fairly quickly TiVo and Replay became, in some sense, redundant. The box had a life all its

THE REVOLT OF THE MASSES

own. The black box was not, like the VCR, a winner-take-all market. There was room for a lot of different companies to sell the same seditious technology and to coexist happily with one another. The box was perfectly designed to seize control of a $50 billion industry from its creators; there was more than enough booty to go around. The only question was: did the TV industry's central authorities have the power to prevent the technology from spreading? "The one question our investors did ask us," said TiVo's CEO, Mike Ramsay, "is 'How long will it take for the TV networks to hate you so much that they shut you down?'"

A good question. If you talked to enough people at TiVo and Replay and pestered enough people at the U.S. television networks and the big advertising firms, you came to realize that they had two stories to tell: an official story and a true story. The official story, even in early 2000, when the boxes went on sale formally, was believed by practically no one, not even journalists. It was pure ritual, made necessary by the desire of everyone concerned not to contemplate the violence about to occur inside a huge industry. The official story was that these new black boxes would not destroy the television industry as we know it and undermine the businesses of all those who depend on the television to sell their products. They'd merely enable the television industry's current rulers to make the industry an even better place.

The official story nicely illustrated the etiquette of the new dialogue between outsiders and insiders. So long as they agreed to keep up appearances, the rule breakers enjoyed a new kind of access to the offices of the rule makers. In all of their meetings with television and advertising executives TiVo rather absurdly represented itself as the soul of corpo-

rate responsibility, mainly because Ramsay and Barton wanted to avoid the sort of expensive lawsuits that ruined Napster. They presented themselves to the networks not as a
threat but as a wonderful business opportunity. To pull this
off, of course, they had to play down a lot of what made their
box desirable to a consumer. Instead of a button that explicitly enabled the viewer to skip commercials, for instance,
Barton designed a technically inelegant but diplomatically
ingenious fast-forward button with three speeds, which
might be called fast forward, faster forward, and faster-faster
forward. The TiVo user was able to pass through the commercials at blinding speed but not skip them entirely: the ad
still made some sort of blurry impression on him. "Network
psychology is to have a line in the sand mentality," explained
Ramsay. "If you're on one side of the line, you're their
friend. If you're on the other side of the line, you're their
enemy. Advertising the ability to skip commercials is on the
other side of the line. We designed the technology so that it
doesn't infuriate the networks."

That was just the polite way of telling someone that you
intended to destroy his business. Replay Networks took the
other approach. Replay, soon renamed ReplayTV, at first
took the position that the networks' interests were irrelevant. What the American consumer wanted, the American
consumer eventually got, and if you failed to give it to him
right away you risked losing his patronage to someone who
would. Replay's remote control has a button marked "Quick-
Skip," which enabled the viewer to leap ahead in increments
of thirty seconds, the length of a typical television commercial. The owner of the Replay box was thus the open adver-
sary of the television establishment. "I spent a lot of the first

year getting thrown out of meetings at networks," Anthony Wood admitted, with some pleasure. Then, tellingly, came a change of the Replay heart. It coincided with the astonishing realization that if the company was willing to lie about its intentions, the establishment might actually help them to succeed. Wood was replaced as CEO by Kim LeMasters, the former president of CBS Entertainment, who saw the point of network support. LeMasters struck a much more conciliatory note. Though he wasn't able to scrap QuickSkip, he let it be known that he would not promote the feature, and that ReplayTV saw itself as the symbiotic type. "The Replay device doesn't do any good if it doesn't have anything to broadcast," LeMasters said.

And so now the two companies—which were nothing more than handfuls of outsiders with a bit of venture capital—were in roughly the same position. Both were arguing to the giant U.S. television networks that a device that stole power from the networks and handed it to consumers was actually good for them. They offered two points to support the case. The first was that the television viewer is too inert for the television ever to change. Several times since the first commercial broadcast in 1939, a new accessory has appeared that promised a revolution—the VCR, the remote control, cable TV—only to be assimilated without greatly disrupting the existing social order. The VCR had proved too unwieldy to be used for anything but rented videos. The remote control enabled people to surf, but not so much that they spooked Procter & Gamble and General Motors and the rest. Cable TV fractured the mass audience into slightly smaller pieces, but again without a huge effect on the economics of the business. True, the big three U.S. television networks had 91 percent of the viewing audience in 1978 and only 45 percent in 1999. But it is also true that of the $45 billion of television advertising in 1999, $14 billion went to CBS, ABC, and NBC, which was $10 billion more than they collected in 1978. (Advertisers have proved willing to pay the networks more for less. It's as if what matters to them is not the absolute size of an audience but the relative one, and the three major networks still offer them the biggest.)

The other point is that by making television more appealing, the black box encouraged people to watch even more of it. This prospect may cast doubts on the future of intelligent life, but it should, in theory, be good for people who currently make and sell TV programs. Replay now had actual data to prove that its new customers watched, on average, three hours more television each week than they did before they got the box. "Yes, we're messing with your business," they argued to the networks. "But in the end, you'll love us for it because three more hours a week means billions for you in additional advertising revenues." Marc Andreessen, for one, believed that this argument was persuasive to networks. "They want to believe it because they are seeing data for the first time that shows young people are watching less and less TV and spending that time on the Internet."

That was the official story. It was the story that enabled TiVo and Replay employees to interact pleasantly with network and advertising executives, as it left the insiders with their dignity. But as I say, no one could possibly have believed it, and it became less plausible every day thanks to the information piling up inside TiVo and Replay about how ordinary people were already using their new black boxes.

They were using them to undermine, with ruthless precision, the interests of TV networks and mass-market advertisers. The owners of the 400,000 or so black boxes installed by the spring of 2001 exhibited two distinctly unsettling new habits.

The first was that they didn't watch scheduled TV anymore. According to Josh Bernoff, a television industry analyst with Forrester Research in Cambridge, Massachusetts, who closely followed both black-box companies, viewers "get into the habit of not paying attention to when the programs are on and just watch what they've recorded."

Well. If it didn't matter when programs ran, the whole concept of prime time vanished, and with it the network's ability to attract an audience for a new show simply by broadcasting it when people have the tube switched on. With it also vanished the special market value of prime time—though the market value of other broadcast space rose. Ditto the idea of pitting one show against another by virtue of its time slot. In the age of black boxes, every show ever broadcast competes against every other show for the viewer's attention. For this reason, whatever advantage a network has in the development of new TV shows disappears. The TV schedule goes from being pyramid-shaped to pancake-shaped. Fringe programs have as much of a chance to attract attention as programs backed by the central authority. The effect of the shift in power toward fringe programs would be to encourage talented people to make them, rather than the mainstream ones. Every Doodie would have its day.

But that wasn't the worst news that TiVo and Replay had for the television networks. The worst news was that hardly anyone who owned one of the boxes watched commercials anymore. Eighty-eight percent—88 percent!—of the advertisements in the programs seen by viewers on their black boxes went unwatched. If no one watched commercials, there was no commercial television.

And yet—and here was the punch line—the major broadcast networks did nothing but encourage the new technology. In August 1999 Time Warner, Disney, and NBC, among others, sank $57 million into Replay. About the same time, NBC and CBS, among others, handed $45 million to TiVo. By the end of 1999 all three major television networks, along with most of the major Hollywood studios, the two biggest Hollywood talent agencies (ICM and CAA), and all the major cable and satellite TV companies, had either made investments or formed partnerships with both Replay and TiVo. There were many reasonable-sounding explanations as to why the networks had rushed to embrace their own creative destruction, most of them premised on the idiocy of network executives. Only one of these explanations was plausible: the central authorities felt they had no choice but to cede power to the fringe. "If the networks could roll back the clock and prevent digital technology from ever happening, they'd do it," said LeMasters of Replay. "But how do you stop progress? We're offering them the chance to adapt."

Tom Rogers, the former president of NBC's cable division who made the first network investment in TiVo and Replay, put it this way: "We thought that the technology was going to come, and it was better to have some voice in shaping it than none." It was the promise of NBC's imprimatur, in fact, that led TiVo to design its remote control without what Rogers called the "ad zapper." "We couldn't be in a position of being seen to promote a technology that was intended to undermine the economic support of the industry," said Rogers. But that was just a fig leaf, and wound up, as fig leaves do, on the ground. By the time Replay decided that it might be useful to have the endorsement and money of the networks, the company was too far along to eliminate QuickSkip, its ad zapper. The company promised not to market the feature; that was enough to persuade NBC to give Replay money. That was in late 1998. Since then Rogers has left NBC to become chairman of Primedia, a holding company for lots of little niche media. Replay has gone back to marketing its ad zapper. And TiVo has launched its own ad campaign, featuring a network executive being hurled out a window by a pair of TiVo's goons.

Their indiscriminate dumping of money onto both TiVo and Replay suggested that the networks understood that the companies trying to commercialize the technology were merely the vehicles for an unalterable force. (Why not just play one against the other, and back the one who promises to be less hostile to the status quo?) It was the technology that mattered; and it was the technology that was sure to win. "A lot of these guys had their bell rung four years ago by the Internet," Steve Shannon of Replay explained to me, "and they don't want to be humiliated a second time." The Internet had spawned a new corporate religion to replace the one it killed. The religion said: Change is inevitable, so you might as well pretend to love it lest you be taken for a doomed species. The question now being posed by the television establishment—and it emerged from the belly of the beast as a weak burp rather than a loud blast—was no longer, "Is this new gadget going to affect us?" or even "Will this gadget eventually change how Americans watch TV?" but "When this gadget changes how Americans watch television, what else will it change?"

A lot.

The basic formula for making and selling TV programs hasn't changed since the beginning of commercial television. The network that develops a new program assumes it can guarantee a measure of success by placing it in a desirable time slot, when a lot of people happen to be watching TV. It further assumes that it can pay for it by selling commercial time during that program. The commercials then get flung at whoever happens to be watching at the time. The entire history of commercial television suddenly appears to have been a big corporate plot erected, as it has been, on control from above rather than choice from below. The networks have coerced, or attempted to coerce, consumers into watching programs and commercials in which they have no native interest. The advertisers who pay for the commercials agree to believe, on astonishingly weak evidence, that some meaningful percentage of viewers actually watch commercials. They have further agreed to believe, again without good evidence, that the sort of people who watch a particular program have more than an ordinary interest in the products advertised on that program. People who enjoy pro football are more likely than people who watch soap operas to drink beer; therefore beer companies buy ad time in the middle of football games.

Against the backdrop of the Internet this arrangement struck the newer sparks of the ad and marketing business as terribly retro. "The television advertising business," said Tim Hanlon, a media director at Starcom Worldwide, a large advertising and marketing conglomerate, "is a science based on specious data." That data, generated by Nielsen Media Research, uses a sample of five thousand homes to determine how many households tune into a given program, not how many watch the ads. "The measurement we use today is very crude," said Daryl Simm, the former head of worldwide media and programming for Procter & Gamble and the current head of media at Omnicom, yet another large advertising and marketing conglomerate. "It's an average measurement of the number of viewers watching an individual program that does not even measure the commercial break. When you think about improvements in measuring viewer habits, you think not about incremental changes but great leaps."

The data-gathering talents of the black box—of the Internet in general—was another good example of society getting the technology it deserved. The picture of the consumer that producers depended upon to sell them things was not merely crude; it was fuzzier by the day. The picture was created mainly out of demographic data—race, age, sex, religion, zip code—and assumptions about how people with certain demographic characteristics behave. Not very long after it began to falter as a socially acceptable guide for explaining anything about a human being, demographics began to falter as a tool for explaining and predicting consumer behavior. Any generalization based on race, age, sex, and religion (though perhaps not zip code) was widely regarded as a form of bigotry in most of the developed world. Even if you thought it—even if it was true—you couldn't say it, or appear to act on it. Perhaps the growing taboo of stereotyping people by race, age, sex, and religion helped people to free themselves from the behaviors associated with the stereotypes, and so they were becoming, all by themselves, more elusive to those who wished to understand them for commercial purposes. Or perhaps that is to mistake cause for effect. In any case, the old categories were inadequate. Consumers needed to be reunderstood by the market. They needed, in effect, new identities.

How handy, therefore, that the Internet and Internet-related gadgets made it possible to judge people based on other criteria: their behavior. Or, rather, a telling slice of that behavior: how they handle a remote control. The TiVo and Replay boxes accumulated, in atomic detail, a record of who watched what, when they watched, and even *how* they watched. Put the box in all 102 million American homes, and you got a pointillist portrait of the entire American television audience. Which raises the second and more disturbing question for the TV industry: What do you do when you actually know who is watching and why? Already TiVo and Replay know what each of their users does every second, though both companies make a point of saying that they don't actually dig into the data to find out who did what, that they only use it in the aggregate. Whatever. They know.

More to the point, they know, in great detail, the viewer's interests, as recorded by the black box. And the economic value of that knowledge is vast. Even now, advertisers pay a lot more per viewer for a targeted ad—however imprecise the targeting might be—than they do for the sort of near-blind matchmaking that the networks, historically, have made their chief business. Put another way, an audience of 200,000 people you know intimately might be more valuable than an amorphous mass of 20 million. After all, a person with a deep interest in a subject is more likely to watch an ad about that subject. "You and I may not care to watch a commercial for Preparation H," said Josh Bernoff. "But for someone with hemorrhoids, it might be the thing he is most eager to hear about. And he's the one the makers of Preparation H want to talk to."

That is the market promise of the new black box. It can extract far more profit from every viewing minute of American television by creating endless clusters of new and very valuable groups of people with some common commercially exploitable interest. "This technology will encourage all sorts of niche brands," said Jim Barton of TiVo, "as well as whole new markets." His favorite example is the field-hockey channel. Everyone in the world with an interest in field hockey can punch "field hockey" into their box, and the box will go and find and record any program having to do with field hockey. At the moment, there isn't much field hockey out there on the tube; that will change. The maker of the new field-hockey-related shows will rent cheap time—at, say, 4 A.M.—to broadcast. Field-hockey enthusiasts will simply record the shows. And—voilà—a new business is born. "The business is two guys," Barton explained. "One of the guys goes out and acquires field-hockey content. The other guy calls people who make field-hockey equipment."

The economics of targeted ads is so compelling that to make them possible is to make them certain. The formula for a field-hockey channel that sells only field-hockey equipment or a hemorrhoid channel that sells only hemorrhoid treatments is endlessly reproducible. But the same slice-'em-and-dice-'em logic applies even to such seemingly mass-market events as the Super Bowl and the Academy Awards.

THE FUTURE JUST HAPPENED

The broadcaster that owns the rights to a mass-market event will be under tremendous pressure to carve the audience up into little pieces and to sell each piece to the highest bidder. Once the black box is ubiquitous, an advertiser need not buy the whole audience; he can buy a piece of the audience. Of course, General Motors may still buy time during the Super Bowl—and pay a lot more for it. The company will probably use the time differently, though. In a world filled with black boxes, GM might use its thirty seconds to distribute fifty different commercials to fifty different clusters of consumers. New mothers will see ads for SUVs, middle-aged people will see ads for sports cars, and so on, and all the little groups will have been identified for GM by the new black box.

But even that is a retrograde example. The operative unit in TV ratings will no longer be the program but the moment. Advertisers and networks will know with weird accuracy who and what within each program best holds television viewers' attention. The black box can determine which joke in a comedian's monologue prompted certain viewers to switch the channel or which medical emergency inspired viewers to exit *ER*. (If you thought the pressure on entertainers to be perpetually entertaining couldn't increase, think again.) The goal of producers will be to intersect with consumers who are most receptive to their pitches, at the moment of highest receptivity.

This went some way to answering the obvious question posed by the box: How do you get people to watch ads when, with the press of a button, they can eliminate them? You show them ads that they want to see, when they want to see them, in the form most convenient to them. The trick was knowing in advance, in precise detail, what people wanted to see. You only had to spend a few hours at TiVo to see that the possibilities along these new lines were endless. The company employed a young man named John Ghashghai, who had a background in analyzing complex data. He was already exploring the trove of information that came into TiVo from its three hundred thousand or so black boxes and was brimming with insights into the mind of the American couch potato. For instance, he'd discovered that the television viewer was not nearly so sedentary as formerly suspected. The typical viewer of live television clicked some button on his remote control one and a half times every five minutes. The typical viewer of recorded television went to the remote once a minute. Most TiVo users watched recorded television, so the typical TiVo user was sending a little piece of information about his viewing habits once a minute. Already TiVo knew that people who watched football games watched, on average, only ten minutes of football games, and which ten minutes were the most heavily watched. TiVo was also able to ferret out strange correlations. For example, people who watched sitcoms were more likely to skip commercials than people who watched hour-long dramas.

Advertisers would pay dearly for scoops such as these, but really they were just the crude beginning of a new pseudoscience. The new data whetted the appetite for the even richer information that was likely to follow, and the obvious endgame: perfect mutual understanding between producer and consumer. "Take Procter & Gamble or General Motors as examples," said Ghashghai. "Instead of mass-producing some new toothpaste or automobile based on individual studies in small samples—what if they had a perfect snapshot of America, with 50 million people instead of five thousand. The waste that can be eliminated by fitting the product to the market, then getting your product in front of the right person at the right time, instead of just hurling messages indiscriminately into the ether, is enormous."

The notion that the producer could come to know the consumer in whole new ways had been hammered into the business mind by the Internet Boom. The Internet gave rise to fantastic efforts at data collection. Between 1996 and 2000, companies were born to do nothing but track where people went on the Internet and what they did there, so that they could be "profiled" and matched up with people who wanted to sell them things. The CEO of DoubleClick, the most famous of these so-called data-mining companies, had put a fine point on the nature of the fever for consumer data when he said that "personal information is the oil of the twenty-first century." The new black box made the Internet wildcatters look like shallow drillers. If it was valuable to observe the two hours each day that half the population spent on the Internet, imagine how valuable it would be to know what the entire population did with the six hours each day it spent watching television.

To use the consumer's mind as a highway to his wallet sounds invasive, and perhaps it is. But—here's the important point—it's nothing personal. Or, at any rate, there is a way to construe the new highly personal relationship impersonally. "This isn't an invasion of privacy," said John Ghashghai, "this is an enabling thing. Ninety percent of this can be done without knowing who you are—without separating you out from the group you've been identified with. We might find you are a boat lover. But we won't separate you out from the other boat lovers. TiVo the company does not need to know who you are; TiVo the receiver does. So long as TiVo the receiver can send back the anonymous data, there are tremendous gains of efficiency to be had." TiVo or Replay or some black-box service company will be able to present some mass-market company trying desperately to stay alive with 40,000 consumers classified as People Who Live for Onions. The individual consumer need never be mentioned by name or separated from his discrete group of onion obsessives—at least not yet. Permitting himself to be classified with ever more intrusive precision is the price the onion obsessive pays for getting his onions, when he wants them, at a better price. He may not like the way the market classifies him, but he has no one to blame but himself, as his identity is merely a reflection of his personal behavior. In that sense, it's rather heartening.

On the other hand, some people would be upset about the loss of privacy—of course they would! But the sheer financial value of the data that TiVo could collect cast the human desire for privacy in a new light. Privacy is no longer a right but a wasteful luxury. The Internet has not merely suggested new weapons for the Invasion of Privacy. It has created terrifying economic incentives for people to abandon their charming old attachment to their privacy. Privacy is newly inefficient if the larger social goal is to get the most stuff to the most people at the cheapest prices. And who would deny that the consumer demand for ever more stuff at ever cheaper prices is one of the great deterministic forces in history? Any technology that gives the consumer what he wants, when he wants it, at a better price, is likely to succeed, in spite of a lot of objections from hoary old privacy nuts.

Of course, there are cultures on earth famously less enamored of consumer goods and more wedded to privacy than American culture. Too bad for them! Consumerism isn't a luxury; it is the necessary behavior underpinning any successful modern economy. It is one of those horrible American traits that other societies have been adopting because they need to adopt it if they wish to remain competitive. With the possible exception of Margaret Thatcher's Britain, they don't do this consciously, as part of some twelve-step program to commercial success. That is the beauty of these deterministic forces: they just happen. Opting out of the oil-patch approach to personal information isn't really an option, at least not in the long run. Say the famously private British people bridle at the notion of a black box in their living room recording their television viewing habits. Say they even raise a ruckus and forbid companies like TiVo to observe them as they watch television. Companies that sell things to British people—British companies—will have inferior data and waste great sums marketing to them. In the long run, the British consumer will pay the price. He will find himself in the same unfortunate position as the big company—the insider—that failed to adapt to a challenge from the outside.

M any things changed when television was able to whisper finely tuned messages to like-minded consumers rather than hollering crude messages through a bullhorn at millions. One thing that changed was the price of the messages. If they are to become more valuable, the targets must shrink, and as the targets shrink, the tools used to hit them must shrink as well. Not even General Motors can spend $3 million on an ad that will only be seen by 40,000 people. "We sort of see this as the changing of television as a medium," said Tim Hanlon of Starcom. "I know the creative side of our business truly hasn't gotten this yet; they still see it as a fringe technology. But they are the ones who will get steamrolled first and most cleanly."

The sad truth about most popular TV programs is that they are poor vehicles for delivering advertising messages. And the sad truth about ads—even the ones that cost $3 million to make and win the Golden Lion at Cannes—is that the people who watch them really don't ask to see them; people are just too lazy to avoid them. The black box put an end to that racket. It might not mean the end of commercial television, but it most certainly means the end of commercial television as we know it. Either the ads need to become as entertaining as the programs or the programs need to contain the ads, so that they cannot be stripped out. If Jennifer Aniston wants to remain a Friend, she might need to don a T-shirt that says DIET COKE—assuming that the Friends' audience winds up being viewed as one that is particularly receptive to pitches from soft-drink companies.

The people who use the mass market's biggest bullhorn are also in trouble. TiVo's CEO, Mike Ramsay, recalled how in late 1997, just after TiVo opened its doors, he received a call out of the blue from Procter & Gamble's research division. Along with General Motors, P&G is the largest buyer of television time in the United States; between them, the two companies ponied up $3 billion of the $45 billion spent in 1999 on television ads. "These two guys from Procter & Gamble were in a car on a cell phone down the street," Ramsay said. "They were in the valley visiting and heard what we were doing and said they'd been playing with a similar idea in their labs because they knew that, sooner or later, something like this was going to happen. And they had the obvious question, 'How do we sell soap now?'"

The P&G research division believed that the inevitable collision of the computer and the television made it far less likely (a) that people would gather in groups of millions to watch TV shows, and (b) that people would watch ads that were thrust on them unbidden. But in P&G's view, this was not necessarily a bad thing. "I'm really intrigued by this notion that the viewer now will be more dedicated," said Daryl Simm, who until 1999 ran P&G's media the world over. "He'll have a higher degree of interest in what he's watching because he has an investment—he's gone to the trouble to capture the program. That investment is going to connect him to the viewing experience in a way that is stronger than just grazing around. Viewer loyalty has got to translate into advertising opportunities."

It does—but for whom? It's one thing for the Internet to poach a bit of the global attention span from the television. It's another to transform the television into an Internet-like renegade force for individualism. The television, pretty much all by itself, defined the mass market. Without the television, there never would have been Tide or Rice Krispies or Alpo but fifty smaller versions of Tide and Rice Krispies and Alpo. This may not seem like a big deal to a user of Tide or Rice Krispies or Alpo, but to a manufacturer of Tide or Rice Krispies or Alpo it matters very much indeed. For the big brands, life without television is no life at all. Giant corporations whose sole purpose is to mass-market consumer goods exist in their current form because the television shaped the mass market. If television ceases to be a mass market, the mass market largely ceases to exist. The question isn't "How does P&G sell soap?" but "How does P&G survive?" It must transform itself from a maker of mass-market goods into the world's largest boutique. After all, the consumer would obviously prefer not only the message precisely tailored to him but the products as well. In this new market, there will be dozens of versions of Tide.

The mass market is crude and inefficient and ripe for reevaluation because Market Man is, too. The new technology enables the market to redefine the consumer along significantly different lines. What a man does with his remote control may not be a perfect reflection of the contents of his character, but it is a lot more perfect than judging him solely by his simple surface traits—especially when the endgame is his habits as a consumer. Nick Donatiello, the head of a San Francisco market research company called Odyssey, says that the black box—along with related technologies like the Internet—makes it likely that ads will be tailored not to outward characteristics but to the more fundamental attitudes of the consumer. General Motors will run one commercial, perhaps, for people with a tragic view of life and another for people with a comic view of life. "Demographics used to be a good proxy for attitudes," Donatiello says. "In the fifties, you could tell a lot of things about a person if you knew where he lived. You can't do that anymore. We've become too fragmented and autonomous a society."

And so the wildcatting approach to personal information was yet another illustration of the Internet's gift for creating new masks. How do people respond when the market attempts to fit them with a mask? Do they come to see themselves as the market sees them? Have people felt more "29–45" or "male" or "Hispanic" because the incoming commercial signals for the past forty years have been aimed at these specific traits? Will they come to think of themselves not as white or young or female but as Positivists or Relativists or whatever other types get dreamed up in response to the data generated by the black boxes? Stuff like this happens in America and, as a result, outside of America. One paradox of that curiously global mass entity, Generation X, is that it viewed itself as ironically detached from the marketplace when in fact it was itself created by the market. It grew out of MTV, which came into being because advertisers found it handy to have young people stripped out from the rest of us so they might be more accurately targeted.

When a persuasive new technology appears, it is only natural to wonder what effect it might have on the world around it. But it is also worth putting the question the other way around: What effect does the world around it have on technology? Why is it that we find a use for such a tool now? Nick Donatiello makes the point that the black box is ideally suited for American life as it is currently configured, when consumer choice has been exalted to a fetish. "If you had offered Americans this box thirty years ago," he says, "they wouldn't have had the same reaction. One of the reasons people used to watch TV in the 1950s and '60s was for the shared experience. The metaphor for the country was the melting pot: people wanted to be the same. People read

Time and *Newsweek* mainly because other people read *Time* and *Newsweek*. Now the metaphor is the quilt."

This is another way of saying that a technology that was shaped by one kind of society is being forced to adapt to a new kind of society. Most of the changes the black box so grandly encourages are merely extensions of trends under way: decentralization, control at the edge of the network, free agency, the rooting out of all kinds of antimarket behavior, and so on. The mass market is a collective. It is an elaborate network of invisible taxes and subsidies, paid with human attention. People who watch commercials subsidize people who don't; people directly influenced by ads subsidize people who watch ads with ironic detachment. This little pocket of socialism came into being at least in part because the technology did not exist that could measure, and put a price on, the attention of individual consumers. The mass market put a price not on individual states of mind but on the average state of mind of commercially very different people. It did this because it made no economic sense to parse in microscopic detail what each and every one of us did with our attention and why we did it. And so the market just lumped us together and assumed we all paid more or less the same attention. Now, suddenly, the technology has appeared that can unravel the collective, and it has created fantastic incentives for capital to complete the job. That it arrives at a moment when all forms of market socialism are on the run is either a magnificent stroke of luck or a good example of a society getting the technology it deserves. The only question is how far its logic will be taken—to what level of detail will the consumer's state of mind be measured and priced?

Maybe the best way to see what's about to happen to the mass market is to observe what has happened already. There's a moment when the center feels in its bones that the ground beneath its feet is moving, whether it likes it or not. The people upon whom change is inflicted may not be as clever as those who inflict the change, but they need not be viewed as stupid. The center has a way of anticipating ever so slightly the great changes it must soon undergo. To some extent, for instance, ads have become more like entertainment, and TV programming has moved in the direction in which it is about to be shoved much, much further. The few events that really benefit from being watched live—sports and awards and sensational unfolding news—have a greater gravitational pull, and a greater market value, than ever. Synthetic events like *Who Wants to Be a Millionaire* and *Survivor* are prescient, for they involve the viewer as a quasi-participant and require the actual participants to deploy many vendible goods, thereby offering sparkling opportunities for product placement. In a "real" world, real goods and services are more naturally introduced than in a purely fictional one.

Already there's some rumbling in the netherworld of advertising and marketing that suggests it is preparing itself for the coming earthquake. For instance, in late 1999, as the first black boxes landed on store shelves, Starcom began to classify television audiences not by demographics but by something it calls "passion groups," which are defined by shared interests. Odyssey shuns demographics and instead categorizes consumers along the lines of their fundamental attitudes, giving them funny names like New Enthusiasts and Old Liners. Procter & Gamble has created a web site called Reflect.com that enables shoppers to create their own beauty products—a harbinger of an age in which every consumer will feel free to demand products tailored to him and him alone. P&G's brands become more frayed every day (there are nineteen different versions of Pert shampoo). The theme of all this—and much of what is new in the market— is that groups are narrower and defined by interests, and that the ultimate interest is . . . Me! The main thing about Me! is that he is encouraged to behave like a child. He always gets what he wants, or at any rate what he thinks he wants. The mass-market consumer was a character who subjected himself to some form of coercion. The unmassed consumer needs to want to be sold. He has been granted a place on the new fringe. He must remain a willing supplier of information to, not merely a simple recipient of messages from, the center. That is the price the system pays for keeping him trapped inside of it.

There is a pitiless economic process at work, so gradual that it does not really ever demand to be noticed. It is a species of economic determinism, the reverse of the one Marx described. The means of consumption, not the means of production, are the engine of modern economic life. The consumer's neurons will be measured and priced only if the consumer wants his neurons to be measured and priced, because their precise measurement enables others to give him exactly what he wants. If this is a conspiracy, it's a whole new kind of conspiracy. The consumer must conspire for, and also, strangely, against himself.

I t's been a good forty years since intelligent observers first pointed out that political candidates were sold much like any other mass-marketed good. It stands to reason that when the mass market fragments, and goods formerly sold through it find better ways to the market, politicians must follow. If the tools used to open the mind of the consumer have proven less and less effective, it's unlikely that those used to melt the heart of the voter were any better. Once the market discovers a new and better tool to study and manipulate the consumer, politics will necessarily find a way to adapt it to its needs. The astonishing thing about the Internet in this regard was how quickly it happened.

At the start of 2000 a truly weird and possibly inspired company founded by a pair of Stanford political scientists, Norman Nie and Doug Rivers, finished spending tens of millions of dollars to install Web television sets in 40,000 American homes. The company, called Knowledge Networks, was trying to address the single greatest problem in political polling—getting a random sample of Americans to answer questions—by paying a random sample of Americans for their time. At least that's how it started out; it's ambitions soon grew. But back in the summer of 1999 Doug Rivers had sent out 70,000 letters, most of them containing ten-dollar bills. The money was the teaser for the big offer: spend ten minutes each week answering his questions over the Internet, and Rivers would give you a free Web TV, free Internet access, and a raft of prizes doled out in various contests and raffles. If you were uneasy with new gadgets, Doug Rivers promised not only to give you the TV and the Internet access but also to send an engineer to install the stuff. An astonishing 56 percent of the people Rivers set out to contact took the offer—compared with the roughly 15 percent now willing to answer questions from a stranger over the telephone.

The new technology landed in the least likely places—trailer parks, urban ghettos, old folks' homes. And—just like that—it was possible to use the Internet to monitor a representative cross section of the U.S. population.

I'd visited several of the homes equipped by Knowledge Networks. The least-educated people had no trouble opening up their brains to the new mind-sucking instrument. There was only one big exception to the general population's embrace of the new: old people. The elderly were relatively slow to use the Internet, slow to buy new gadgets like TiVo, and hard to persuade to become part of new social experiments such as the Knowledge Networks panel. When I set out to make an example of one of the new guinea pigs I went looking for an older person, thinking that maybe the exception might say something interesting about the rule. And I wound up on Marion Frost's front porch.

When Marion Frost received her first letter from Doug Rivers, she had just turned eighty, which meant she was gold to any pollster looking to build a random sample of Americans. Along with Americans who earn more than $150,000 a year, Americans who have less than an eighth-grade education, and Americans who don't speak English, Americans over seventy-five tend to elude pollsters.

Frost has lived in the same quaint cottage for forty-six years, nestled in a middle-class Silicon Valley neighborhood doing its best to avoid being overrun by property developers. The only hint of frailty about her was the cast on her left wrist, which she had broken, absurdly, on her way back from the DMV, where she had gone to obtain a handicapped parking sticker. The only sign that she found it odd for a complete stranger to show up at her house to watch her watch television was that she had invited a friend over to join us. The three of us—me, Frost, and her friend Yvette Reyes—settled down to a spread of pizza, cookies, and coffee in a living room that doubled as a shrine to bygone values. The furniture would be familiar to anyone who had grandparents in the 1960s; the television was one of those giant oak cabinets with chrome dials that they stopped making back in the 1970s. "My husband died eighteen years ago," Frost said, "and we bought the TV at least fifteen years before that." The single anomaly in the place was the black Web TV box on top of the television cabinet. With its infrared ports, flashing lights, and miniature keyboard, the thing was as incongruous as a Martian.

It was the night of the second Bush-Gore debate, and CBS was using Knowledge Networks—and by extension Marion Frost—to conduct two kinds of polls. One, which Dan Rather was calling "a snap poll," would measure who won the debate. The other, which Dan Rather would never mention, since CBS was still testing it and was terrified of making a fool of itself, would seek to understand why he won. Twenty minutes before the debate, the red light on top of Frost's black box began to flash, its way of saying that it was waiting for Frost to switch on her TV and answer a few questions from Knowledge Networks.

Among the many things I was curious to know was why Rivers had been so successful at luring Americans into being the rats in his massive laboratory experiment. This was the astonishing fact: people unwilling to be questioned over the phone were proving eager to be examined with what was, ultimately, a far more invasive tool. I assumed that in all cases Rivers had appealed to the lab rats' insatiable lust for money and freebies.

"How did these people talk you into taking their surveys?" I asked Frost.

"They just called me out of the blue," she said. She sat toward the edge of her chair, her arm in a dark blue sling. All around her were photographs of children. She had reared three of her own, adopted another, and taken in seven foster children.

"But you got their letter with the ten dollars in it?"

"They never sent me a letter with ten dollars in it," she said. "I got a letter from Doug Rivers but no ten dollars."

Yvette chuckled softly on the sofa beside me. "You come cheap, Marion," she said.

I motioned to her Web TV. "So what do you find you use it for?"

Embarrassment flashed across her kind face. "The truth is I haven't figured out how to use it."

"Then how do you take the surveys?"

"Oh," she said. "I do all the surveys. When the light comes on I call Robert and he comes over and turns it on and feeds in the information."

Robert was Frost's forty-five-year-old son.

Yvette burst out laughing. "It's true! I try to send her e-mail. When I hear back from her it says 'This is Robert, writing on behalf of Marion.'"

"When I get the cast off I think I'll learn how to use it," said Marion Frost, trying and failing to raise her arm.

"But you still answer the surveys," I said.

"Oh yes," she said, brightening. "Mostly all the questions

are about products. Juice. One I remember was the different kinds of juice. They wanted to know what was my opinion of cranberry juice. Would I mix it in with other juices. I don't know. I figure they want to dilute the cranberry juice with other juices. I told them my opinion."

No ten-dollar bill, no interest in the free Internet access, or the Web TV or, for that matter, the endless raffles and contests sponsored by Knowledge Networks. No sense whatsoever that she was part of a great new experiment in paying people so well to answer questions that they'd come back every week to answer more. Still, she was perfectly content to tune in once a week to boost the annual sales of Ocean Spray. "I don't understand," I said. "Why did you agree to be in this survey?"

She was at a loss for an answer, which was okay, since Yvette wasn't. "Because," she said, in a tone that put an end to further questions, *"she's a good person."*

"But you know," I said, "they're supposed to *pay* you for your cooperation."

"They used to do that at the malls," said Yvette, dismissively. "They used to give you money just for telling them how many pieces you wanted on each roll of toilet paper."

On that note, we settled back and waited for Knowledge Networks to begin measuring Marion Frost's opinions.

One curious subplot of the 2000 U.S. presidential campaign was the huge, inexplicable swings in the polls. There were any number of possible reasons for these—inept pollsters, fickle voters—but the most persuasive was the growing reluctance of Americans to take calls from pollsters. In the previous decade, the response rate to telephone polls had fallen from as high as 40 percent to 15 percent. If the 15 percent of the population still willing to be polite to people who interrupted their dinners were representative of the rest, this trend would not be a problem for pollsters. But they weren't—they were, statistically speaking, freaks—and so the trend was a big problem. The picture was growing blurrier by the day. Every year the American voter was growing less knowable. His reluctance to make himself available to pollsters was the political equivalent of his refusal, as a consumer, to behave as the demographics suggested that he should.

Onto this scene the Internet appeared to have arrived just in time. It gave new hope to people who believed that human behavior might be studied and explained scientifically. Internet polling enjoys several obvious advantages over old-fashioned survey techniques: it's potentially more scientific than chasing down people in shopping malls, it's less blatantly intrusive than phoning people at dinner, and it carries video to those polled so that ads, movie trailers, and product designs can be tested directly. But maybe as important as all these combined is the ease with which an Internet pollster can create a new kind of dialogue with the people he polls. In what Doug Rivers calls "a virtual conversation," the pollster with easy, steady access to a cross section of the population can unspool a detailed story about that population's tastes and habits.

But the Internet has, and will have for some good while, one huge disadvantage for pollsters: not quite half the U.S. population uses it. The problem is even worse in Asia and Europe—outside of Scandinavia. The Internet is quickly being adopted but it will still be a long time in every country, with the possible exception of Finland, before a representative sample of the population is on line. In the summer of 1998 Rivers and his Stanford colleague, Nie, both of whom had made distinguished careers studying polling techniques, discovered that they shared an outrage at the sham polls of the "general population" conducted on the Internet. They got to talking about ways that the Internet might be used to poll properly, short of waiting the years it would take for the technology to trickle down. They decided to go out and identify a random sample of Americans and persuade them to go on line, for free.

Of course, it costs a fortune to dole out tens of thousands of Web TVs. So Rivers, who wound up running the business, was forced to neglect his original interest in political polling and acquire an interest in market research. Corporate America spends $5 billion a year for market surveys. Companies pay roughly $2 for every minute that randomly selected Americans spend answering questions posed by people who pester them at dinnertime. The reason you are worth $120 an hour while you scratch yourself and talk on the phone to a pollster is that pretty much anyone with anything to mass-market—packaged goods, media come-ons, financial products—longs for detailed portraits of the consumer. A network of tens of thousands of Web TVs randomly distributed across the population would represent not just a statistical improvement. It would create a new genre of portraiture.

Typically, the relationship between the American Observer and the American Observed has been a one-night stand. A pollster calls and insists on pawing you a bit, and then you never hear from him again. Knowledge Networks was after something more. Its Web TVs would follow the same people, easily and cheaply, and measure not just their responses to surveys but also their behavior on the Internet. And it would be able to divine patterns in that behavior that companies could then exploit. As Knowledge Networks expanded, it would become possible to poll random samples of tiny populations—people who drank expensive tequila, say, or who voted for Pat Buchanan. "Try finding a random sample of Jews by phone," Rivers says. "Jews are two percent of the population. Do you know how many randomly generated phone numbers you need to call to find four hundred Jews?"

Interestingly, all parties to this new and seemingly intrusive relationship shared a financial interest in it becoming ever more intrusive. There might be people like Marion Frost who don't think of their time as money, but they were a dying breed. By enriching the information he mined from his random sample of American minds, Rivers raised the value of those who spent time answering his questions. The more their time was worth, the more goodies they got, and the more goodies they got, the more willing they would be to answer questions.

It took Rivers just three days to raise the first $6 million he needed from Silicon Valley venture capitalists. The VCs could see exactly what he was driving at; after all, they were themselves social theorists. The only question they asked him was why he was sending ten dollars to people *before* they agreed to his deal. "The VCs said things like, 'If you sent that to me, I'd just keep the money and not do the surveys,'" said Rivers. "I had to tell them that most people weren't like venture capitalists." Once Rivers had proved to his backers that his system worked in a few hundred homes, he went back and asked for another $36 million so he could install many, many more Web TVs. The VCs promptly handed that over too.

By the spring of 2001, 100,000 Americans were spending ten minutes each week answering Rivers's questions, often e-mailing him with extra ideas and comments and news about their lives. ("Mr. Rivers," wrote one, "Terrence cannot answer the questions this week because he is in jail.") By the end of the summer the system was running at full capacity. If Rivers wanted to ask more questions he needed to enlist more lab rats to answer them. So he went back once again to the venture capitalists for as much as $60 million more to install another 60,000 black boxes. The stock market had collapsed, and the venture capitalists were as tentative as they had ever been, but after a bit of hemming and hawing they gave it to him—thus demonstrating the effect of the capital markets on technical progress. The markets can slow progress down a bit, but they have no effect on its direction. By the end of 2002, 250,000 Americans will be engaged in Knowledge Networks' virtual conversation—the fastest, biggest, and quite possibly most accurate tracking poll ever conducted. Rivers had calculated that the global market might eventually demand that the conversation grow to include three million people. Each person would remain in the sample for three years, at which point they are considered too overexposed to polls to be accurately polled.

Even by the standards set by the Internet Boom, this is a fairly sensational financial story. Among other things, it tells you a great deal about what might be called the public opinion of public opinion. It has become more self-conscious.

Between 1876, when the last new polling technology, the telephone, was invented, and the 1960s, when the phone was sufficiently widespread to allow for random sampling, it occurred to no one to go out and install 100,000 telephones in American homes. The telephone didn't become a polling device until it had spread on its own into every nook and cranny of American life. But the world no longer is willing to wait for a more accurate self-portrait. It wants its new identities and it wants them now. Offered even the slightest chance to, as Rivers puts it, "get inside people's minds and find out what's there," investors have proved willing to pay whatever it cost. They do this because they know that the larger economy will pay more than whatever it costs.

The ever-evolving relationship between American consumers and producers inevitably spills over into American politics, which is why a Stanford political scientist has wound up, at least for the moment, testing cranberry juice cocktails for a living. A better view of the public opinion of juice soon becomes a better view of the public opinion of issues and ads and phrases and candidates. Rivers knew it wouldn't be long before some enterprising political consultant used his device enter the mind of the American voter. "But the thing we've found," he said, "is that the political people are slower on the uptake than the businesspeople. In part, it's because they don't have the same money to spend. But it's also that the sort of people who become pollsters to presidential campaigns don't like to hear the answers to honest polls. They're believers in a cause." He was able to say this with detached amusement rather than despair because he knew that the political people had to come around—and how could they

not? Politics is a competitive market. Better polls give politicians who follow them an edge. Those who don't will wind
up being put out of business by those who do.

P eople who bother to imagine how the Internet might
change democracy usually assume it will take power
away from politicians and give it to the people. It's easy to
see how the Internet might lead inexorably to the same
extreme form of democracy that has evolved in California,
where the big issues often are put directly to the people for
a vote. Sooner or later, it will be possible to vote on line. And
sooner or later, it will be possible to collect signatures on
line. Together, these changes might well lead to direct
democracy, at least in states like California where citizens
can call votes on an issue simply by gathering enough signatures on its behalf. At which point someone asks, "Why can't
we do the same thing in Washington?" One constitutional
amendment later and—poof—Americans are voting directly
to decide important national questions rather than voting
for politicians and leaving the decisions up to them.

This line of futurology has history on its side. The original belief that representative democracy is inherently supe-
rior to direct democracy has been dying a slow death, right
alongside the belief in the existence of elites. Every step
taken by American democracy has been in an egalitarian
direction. The direct election of U.S. senators, the extension
of the vote to blacks, women, and adolescents, the adoption
of initiative and referendum in the vast majority of states,
the rise of public opinion polling—all of this pushes democracy in the same direction. It forces politicians to be more

informed of, and responsive to, majority opinion. It nudges American democracy ever so slightly away from its original elitist conception (Alexander Hamilton referred to public opinion as "the Beast") and moves it toward something else. The Knowledge Networks poll offers a glimpse of what that something else might be, a world in which politicians become so well informed about public opinion that there is no need for direct democracy.

It was with something like this in mind that George Gallup began his campaign in the 1930s to make political polling scientifically respectable. Gallup thought that democracy worked better the better informed politicians were of majority opinion. Rivers does not exactly share this view. He created Knowledge Networks because he believed that inaccurate polls are a danger to democracy and an insult to good social science—but that is a long way from Gallup's original utopian vision. Rivers says he believes that Internet polling is inevitable, so that it might as well be done honestly. But deep down he believes that his faster and cheaper opinion-gathering machine will provide politicians with a more detailed snapshot of public opinion, and thus give rise to an even more constipated politics. The more per-fectly informed politicians are about public opinion, the more they are chained to it. As Rivers puts it, "The problem right now isn't that politicians in Washington are out of touch. The problem is that they're too closely in touch. And this will make the problem worse." In short, you may believe that politicians could not be more automated than they are now. Just wait.

But it isn't only the politicians who are changed by the technology. The more perfectly watched are the voters, the less they have to pay attention to politics. After all, there's no point in anyone but a revolutionary participating in a system of majority rule when the will of the majority is always, and automatically, known.

Business has a talent for inducing change that democratic politics lacks, as politics must persuade a majority of the electorate to abandon its old habits before anyone can budge. And it takes a while for an entire culture to get used to the idea that there is no point in participating in democracy unless you are paid to do so. It takes even longer for it to figure out that its participation is worth two bucks a minute. For new technologies to win they must vanquish old ideas; and for old ideas to die, the people who hold them must die first. And Marion Frost wasn't quite ready for that.

Fifteen minutes before the second presidential debate began, her doorbell rang. It was a young man from Knowledge Networks, who had driven an hour to switch on her Web TV. (Frost's son was traveling, and I couldn't figure it out.) The screen, previously given over to Dan Rather's face, went blue. Onto it came a message: "Please try to have fun while being as serious about this test as possible."

It went on to ask several long, pro forma questions, which Frost insisted on reading aloud before turning her attention to the alien keyboard.

Yvette sighed. "This is going to be a long night, Marion."

This was my cue to take her son's place at the keyboard. When we had finished with their questions, the picture came back on the screen, with a long measuring rod at the bottom of it. The rod had a plus sign on one end and a minus sign on the other. Frost was meant to signal what she thought of whatever Bush and Gore said, as they said it, by moving a tiny rectangle back and forth between the two. Instead she told me what she thought, and I moved the rectangle for her. Her stream of opinions would flow into a river through Knowledge Networks' computers and into CBS studios in New York.

The debate started. I waited for Marion Frost's first command. "I like that Jim Lehrer," she said.

Lehrer was the television journalist moderating the debate. He had asked Bush a question about foreign policy, and Bush talked for as long as he could on the subject, then did his best to think up some more words to fill the time. Frost said nothing. The little rectangle didn't budge.

"I don't know about Bush," she finally said, "but I'm glad Jim Lehrer's going to be there."

Al Gore then went off on his usual relentless quest for a gold star, and Frost listened to all of what he said intently, but again failed to respond. She seemed to want to think about what he said, but the new technology didn't want thought. It wanted quick. It wanted impulses. Thumbs-up, thumbs-down.

"I don't know," she finally said, as Bush took over. "I'm confused. I think they're both right on some areas." She was growing ever so slightly distressed at her inability to give the black box what it wanted. Finally Bush said something that caused Frost to say: "I like that. Go ahead and make it positive." But it was as much out of a concern for the little rectangle than actual deep feeling. In any case, her reflex was too slow to hit its mark; by the time I'd moved the rectangle, Gore had again butted in. This didn't seem to bother Frost.

She was too busy trying to make sense out of the arguments Bush had made about the IMF. "That's the International Monetary Fund," she said—for my benefit, I think.

Yvette sighed and headed for the kitchen. "I get to take a break," she said. "You two can't move."

Frost looked at me with concern and asked, "Would you like a cookie?"

The debate heated up again. Gore began to attack Bush's record on health care. At one point he said of his opponent, "I believe he has a good heart . . . however." Frost became irritated. Hearts were something she knew about. "'I believe he has a good heart,' what kind of statement is that?" she said. The implications for the rectangle were unclear.

"Should I make it negative?"

"A little," she said.

On this went for an hour and a half, much like the debate itself, defying any possibility of the reflection or deliberation that Frost was intent on supplying. She watched without much interest Dan Rather announce that Bush had won the snap poll—52 percent to 48 percent. She wasn't by nature impulsive; and so she wasn't by nature well designed to be observed; and so she wasn't by nature easy to fit into the new mask.

The joy of watching Marion Frost watching her Web TV was her insistence on layering old and dying habits of mind onto the new, supercharged process. Her opinions were being monitored as closely as political opinions have ever been monitored, and yet she didn't really allow the monitoring to interfere with her idea of how to watch a political debate. She avoided making snap judgments just as she had somehow avoided getting paid for offering them. She just did what she did because she entertained some notion of her civic obligations above and beyond her economic interests. Either that or she simply could not believe that a citizen is meant to be paid for her services.

FOUR

THE UNABOMBER HAD

A POINT

As I left Marion Frost's house it occurred to me that Knowledge Networks and I had something in common: we both went around bothering strangers. We both wanted to sit in their living rooms and ask them questions and watch them do whatever they did when they thought they weren't being watched. And we both had been greeted mainly with warmth and encouragement. I haven't told you about all of the people whose lives I briefly disrupted because it would mean telling you the same story over and over again. Instead I've selected the few that fairly represent the whole. And I can see that by amplifying a few voices, I may have muted the chorus. Let the chorus sing and it sang with one voice. Everyone I encountered accepted, without a second thought, the idea that the Internet was a necessary instrument of progress. In short, they had embraced the toolmaker's view of the world. The Internet Boom, generally, had enabled the technologist's notion of progress to take root in all sorts of unlikely places.

Ennis, Ireland, for example. I spent several strange days in Ennis watching a lot of charming old Irishmen trying to transform themselves into computer geeks. It was an unlikely place for such a thing to occur. Until 1997, unless you had a special interest in Irish folk music, Ennis was mainly a place to drive through on your way from the Shannon airport to a golf course. Ennis was one of many small Irish villages that feel as if it were born old and poor, and then proceed to grow older and poorer. That changed on April 3, 1997, when the Irish phone company Eircom announced a competition open to towns in Ireland with populations under thirty thousand, to become the officially designated Information Age Town. Each household in the winning town was to receive a free computer and Internet access.

It was never very clear to anyone how to compete for what amounted to a $15 million prize, other than to express a general enthusiasm for the Internet and the Irish phone company. So that's what Ennis did. A small group of highly energetic and patriotic locals, certain that Ennis must die if it did not change, worked the town into a frenzy over the Internet. They held rallies to cheer the Information Age. They made videos of themselves celebrating the future. They taught a thousand schoolchildren to march across the football field spelling out the phrase INFORMATION AGE TOWN, and another group of several hundred to form the shape of Eircom's logo, with the idea that they might coax Eircom's judges into a helicopter on a beautiful summer's day and fly them over the football field. And they did—though there had been a small hitch. On the appointed day the sun was up and the judges were late, and the kids who had been assigned to march in the shape of the Irish phone company's logo began to faint from the heat. This deterred no one. When one kid passed out, they hauled him off and put another kid in his place.

In the end Ennis had won the Eircom competition—but that wasn't the interesting bit of the story. The interesting part was this: right up until they were given free access to it, no one who lived in Ennis knew what the Internet was. They just knew it was something you were supposed to want to have, and that it enabled any town that had it to be a town of the future rather than a town of the past.

The overnight journey of Ennis from medieval Irish village to the Information Age Town had had some strange effects. One of these you can probably guess: it inverted long-standing relations between old people and young people. The kids took to the new machines much more quickly than the adults. All of a sudden the children of Ennis were developing skills and attitudes and ambitions for the global marketplace, along with a keen awareness of the spirit of the marketplace. The old people mainly kept on being charming provincials. I met one elderly fellow who didn't even unpack his free computer, but simply shipped it on to his daughter in London. Others treated the machines as they might a football trophy and propped them up on the mantelpiece, an inert testament to their superiority over rival towns. But, of course, many of the codgers tried to do what was expected of them and reinvent themselves as Information Age warriors. And, of course, to do this they were utterly dependent on their children and grandchildren. To ease the transition, the local school established a series of Internet classes for senior citizens, taught by some of the same children who, a few months before, had marched around the football field in the shape of a corporate logo.

Ennis was just one of many examples of the ground gained during the Internet Boom by the tech-centric view of the world. Paul Romer, the economist at Stanford University who had been the first to prove to the satisfaction of his fellow economists the primary importance of technical change in economic growth, spent a lot of time thinking about the social implications of his work. The Internet, Romer believed, was just an example of a far more powerful force: ever faster technical progress. Technical change was certain to become more rapid because everyone now understood that it was the source of new wealth. And rapid technical change undermined a lot more than just old technology. "Much of our society still retains a patriarchal flavor," said Romer, "in that it is premised on the authority of the elders, and the notion that respect and compensation should improve over the course of your lifetime. It may be that we are moving to a model where the peak earning years occur before a person is thirty years old, after which he effectively retires. It's the pro athlete model, extended to everyone." All sorts of thoughts followed from that one. For instance, what became of the notion of building a career, or taking on projects that required a long time to complete? "In the past you might have thought of making a contribution through something that was truly permanent," said Romer. "You build the pyramid, it'll be there forever. Now, what permanence means is to be a stepping-stone that someone else uses, and that they build on in the future. So it's not something which itself lasts, but it's one more step which others will build from and keep a process going."

The efficient pursuit of prosperity demanded the end of the idea of a permanent entrenched elite. There could be only talented, and usually young, people who came along, gave the world a shove forward, and then exited stage left. Ambitious people needed to understand that their moment in the spotlight would likely be brief. They'd enjoy a green flash of high status, after which they'd be asked to step aside. This was true not just of ordinary people but also of the technologists at the heart of the new order. The people thrust into the spotlight by the Internet needed to understand that their status, too, was ephemeral. The price of being allowed to change the world at eighteen was being viewed as washed-up at forty.

Awareness that he is likely to be washed-up before he feels washed-up can have a strange effect on a forty-year-old, even when he is a technologist. As the Internet went boom there were signs that the new recipe for economic success was not necessarily a recipe for grown-up happiness. The notion that adulthood is, in some sense, obsolete took some getting used to. And a few of the people who had exhibited the greatest faith in technology, and who had done the most to promote rapid change, began to have second thoughts about the whole enterprise. The pistons in their souls began to misfire in strange and wonderful ways.

In 1995 a computer scientist named Danny Hillis wrote an essay for *Wired* magazine called "The Millennium Clock." Hillis was one of those people whose name, when mentioned, inspired other members of the techno-elite to nod and say, "He's smart." Hillis had made his reputation in computer circles in the late 1970s, as a young man in his twenties, when he had designed the world's fastest supercomputer. People who worked on making fast computers even faster were, until recently, the true elite of the techno-class. The faster the computer, the higher the status of the person who worked on it. That was true right up until the moment the Internet went boom, and the speed of the central computer was supplanted as a status symbol by the speed at which one delivered information to the edge, and one's ability to foresee the future uses to which superfast information might be put.

In the mid-1990s, as he approached his fortieth birthday, Hillis became troubled by something he noticed in the people around him. "I was working on making the fastest machines in the world," he said. "And all my customers were worried about faster, faster, faster. And I realized that all this obsession with speed was making them forget about the future. Their time frame was the next nanosecond, not even what's happening next year or next century. And I felt like people were really missing something."

In his essay Hillis complained specifically about the way the future kept closing in on *him.* "When I was a kid three decades ago," he wrote, "the future was a long way off—so was the turn of millennium. Dates like 1984 and 2001 were comfortably remote. But the funny thing is, that in all these years, the future that people think about has not moved past the millennium. It's as if the future has been shrinking one year, per year, for my entire life. 2005 is still too far away to plan for and 2030 is too far away to even think about. Why bother making plans when everything will change?" The awareness that something big had changed, and that the change wasn't necessarily for the better, struck Hillis on a trip to the Old World. He'd visited New College, Oxford, built in 1386. There he'd been shown the giant oak beams in the ceiling of the common room. Unlike most of the building, the oak beams weren't original; they'd been replaced at the end of the nineteenth century. When the time had come to replace the giant old oak beams, giant new oak beams had been thin on the ground. The New College people had called the Oxford forester to explain their problem, and the forester had informed the New College people that the man who had built the original ceiling back in the fourteenth century had also planted the trees to replace the beams. They stood on Oxford University land, waiting to be cut down.

That story had made a very great impression on Hillis. He didn't know whether it was true or not and he didn't even bother to find out. It was one of those stories that *should* be true. More important, it got Hillis thinking: no one he knew had that kind of foresight. His high-tech friends talked incessantly about the future, but what they really meant when they said "future" was "now" or, at best, "next." They were entirely focused on the economic consequences of their next few moves on the high-tech chessboard. No one was thinking about the distant future. "The deep future," as Hillis dubbed it. The more he thought about this, the more his observation troubled him. Why not steal a page from the Old World playbook and build something designed to last a very long time? Something that would encourage other people to expand their horizons.

The idea Hillis finally came up with was to build a clock that would last for ten thousand years. The second hand would tick once a year, the minute hand would move once a century, and the cuckoo would emerge once a millennium. It would serve as a symbol of the importance of the future and persuade people to think of what they might do and think to serve it. He tried out the idea on people he admired. One of them was the virologist Jonas Salk. Just before Salk's death in 1995, Hillis sat next to him at a dinner party and told him about the ten-thousand-year clock. Salk didn't understand it. Hillis tried to explain; he told Salk about the oak beams in the common room at New College, Oxford. "Oh I see," said Salk. "You want to preserve something of yourself, just as I am preserving something of myself by having this conversation with you."

The poignancy of that exchange, and the poignancy of Salk's death a few weeks later, led Hillis to end his essay on a Whitmanesque note of reckless abandon. "People of the future," he declared, "here is the part of me that I want to preserve, and maybe the clock is my way of explaining it to you: I cannot imagine the future but I care about it. I know I am part of a story that starts long before I can remember and continues long beyond when anyone will remember me. I sense that I am alive at a time of important change, and I feel a responsibility to make sure that the change comes out well. I plant my acorns knowing that I will never live to harvest the oaks."

Well. You can imagine the response back at the computer center. Frozen smiles! But rigor mortis of the lips was accompanied in many a geek mind by a terrifying thought: *What if Danny's right? What if I am utterly lacking in vision?* "Shortsighted" is techno-elite shorthand for "doomed"; lack of vision was just something you didn't want to be accused of. Hillis wrote that when he told his friends about his idea for a clock that would tick for ten thousand years, "either they get it or they don't." But enough of them got it—or pretended to get it—that Hillis was able to raise $50 million, buy a mountain in the desert of eastern Nevada into which he planned to insert the Clock of the Long Now (as the clock was now called), and to create a foundation inside a genteel brick building in San Francisco's Presidio, on the board of which sat many of the leading figures of the Internet Revolution.

The building soon was staffed with young and eager employees, who set about turning the clock into a legend. From the start Hillis was clear that he intended to create a world-famous monument that would change the way human beings thought about the future. And so he was also clear that he intended to drum up as much publicity for his clock as he could get. The clock-priests at Hillis's foundation soon generated the most astonishing press releases:

> It has been nearly 10,000 years since the end of the last Ice Age and the emergence of modern civilization. Progress during that time was often measured on a faster/cheaper scale. The Long Now Foundation seeks to promote slower/better thinking and to focus on our collective creativity for the next 10,000 years. Civilization is revving itself into a pathologically short attention span.

There was a lot more along these lines. In their literature the new clock-priests changed the dating system so that, for instance, Hillis's original idea for the clock was said to have occurred to him in *01993*. "Not Stonehenge," a spokesman would say with a straight face, when you asked him which among the world's ancient monuments he considered an adequate model for the Clock of the Long Now. "Stonehenge is a failed monument. Nobody knows how it worked, and the people that were in charge of it are long gone. Even the Pyramids have a problem, in that they were basically ripped off, both interior and exterior, and all they have left is the form. Plus the whole pharaoh approach to life is gone, we have to reconstruct it." There were only two monuments on the planet that the Clock people viewed as even moderately successful models: Big Ben in London and the Ise shrine, which was someplace in Japan. The Ise shrine was a nice touch. It gave the whole Clock enterprise a Mysteries of the Orient feel to it.

The ambition was wildly grandiose. But what made it so special . . . what sent a little shiver down your spine just to think of it was how . . . *down to earth* the people behind it were. The clock-priests at the foundation wore shorts to work. Hillis himself, in the flesh as opposed to on the page, came across as the soul of modesty and self-effacement. Just another guy in Banana Republic trousers trying to change human history. He worked out of an office park in Los Angeles, where he had holed up after he left Walt Disney. That's where he'd worked in between designing the world's fastest computer and thinking up plans for the deep future. He'd been something called a Disney Fellow and had designed cool new rides for the Disney theme parks. The writer Po Bronson had spent a day in Disneyland with Hillis back then and come away with the impression that the new Disney Fellow was a child in a man's body.

And it was true, there was something unformed about Danny Hillis. Five minutes after you left him you had trouble remembering what he looked like. I myself had that experience. I flew to Los Angeles and drove to his office park. I met and spoke to a grown man in his late forties. But the mental picture I had of him faded almost instantly after I'd left him. Whatever he actually looked like melted into a precocious, sloppily dressed fourteen-year-old boy with chubby cheeks and red curls. "Danny" from the television show *The Partridge Family*.

That was okay since it was what Hillis thought, not how he looked, that was important. What he thought was that our pictures of the future from here on out were likely to come pretty much entirely from technologists. "In some sense," he said, "in the world we've created today, technologists are the only ones that can think about the future. It's not that everybody isn't interested in the future, it's that if you imagine that we're going to have the technologies for cloning people and creating designer babies and building artificial intelligent computers and anthropomorphic robots and transporters and spaceships and so on, then thinking about the future is thinking about technology."

Enter the Clock. On the night of December 31, 1999, Hillis and his growing band of techno-elites unveiled a prototype (now on display in London's Science Museum). The design, oddly enough, was a throwback. Although it had been built to take a lickin' and keep on tickin', it looked an awful lot like a grandfather clock. To keep running it required someone to wind it once every eight days—the spiel about ten-thousand years was pure salesmanship. The need for an actual human being to be present to twist the dials got Hillis thinking along new lines. There would have to be an *institution* to employ the clock-keepers. The Institute of the Clock of the Long Now. The Institute would not only support the clock-keepers but sustain a clock library—which of course raised the question of what exactly would go into the library, a subject that was likely to lead to petty squabbling. In a brilliant example of the techno-elites' gift for turning political questions into engineering ones, they put that thorny issue to one side and instead addressed the *technical* problem of how to preserve information for ten thousand years. After all, what was to be preserved was obviously a less pressing question than the manner in which it would be preserved. (And the people who solved that first problem would no doubt enjoy a great moral authority when it came time to decide the second.)

Danny Hillis offered the first hint of an answer to the question: How does an inherently unstable elite respond to the perpetual threat posed to it by the system that gave rise to it? Not well! In fact they get up to the same tricks as old-fashioned important people to insulate themselves from the vicissitudes of fortune. They set out to make themselves into a ruling class. In doing this they used time as their weapon, just as ruling classes have always done. You only had to visit any of the hundreds of recently gelded British hereditary peers and tour the walls of the estate and take in the family portraits and coats of arms and the mausoleums stuffed with dead ancestors and so on to see how heavily their ancestors relied on the past to buttress their authority—and to realize how irrelevant this same past is today for the purposes of gaining power. A lord was a lord because his father was a lord. If a lord is a lord no more but in name, it is because the past has been neutered. No member of the transient elite would be so foolish as to build a monument to the past as he knows such a monument (a) cannot last and, worse,

(b) proves that he has failed to understand that the past is beside the point. In a rapidly changing world in which the past is forever being plowed under, the future is the only game in town. Which is why the internal controversies of the techno-elite typically involve warring versions of the future. And why those among them who can find publishers write books with titles like *The Road Ahead*. (Or maybe *Next*.)

How clever, then, to rethink the purpose of monuments and use them instead to colonize the future. Hillis had said that his goal was modest—to "give people permission to think ten thousand years into the future." But from that modest thought followed a less modest one: henceforth no one would be allowed to think of the next ten thousand years without also thinking of Danny Hillis. He'd written his name in bold letters across the next ten millennia.

The Clock of the Long Now was an inversion of the old Top Dog's instinct for self-preservation. But the ultimate purpose of the monument hadn't changed a bit: to suppress the revolutionary forces that raised the dog to the top in the first place. The Clock of the Long Now was meant to cause progress to pause and think a bit before it took its next step—the step that might well squash some middle-aged member of the techno-elite. The thinking behind it actually had nothing in common with the thinking that led some humble English gardener in 1386 to plant an acorn. It had made a lot of sense back in 1386 to plant a tree that wouldn't be needed until 1886. The man who did the planting enjoyed the luxury of believing, quite reasonably, that the future might be much like the present. His action did not represent some long-lost talent for futurology, or geological feeling for time, but a human assumption that the future wasn't something you needed to think very hard about, because it wouldn't be much different from the here and now.

At about the time Danny Hillis unveiled the prototype of his clock, another member of the techno-elite also well past his fortieth birthday was beginning to register his own protest. His name was Bill Joy and he was the chief scientist at Sun Microsystems. *Fortune* magazine had called Joy "the Edison of the Internet" because he, or at any rate his company, had developed the programming language Java, which was one of the Internet's lingua francas. He—or his company—was also especially quick to sense the possibilities on the Internet fringe and had, for example, gobbled up what they could of the Gnutella successors. Right up to the moment in early 2000 when he flipped on his word processor to write an essay for the April issue of *Wired* magazine, Bill Joy was the embodiment of techno-centric progress. He called his essay "Why the Future Doesn't Need Us."

It began slowly, with a statement of Joy's general affection for his chosen field, computer science. He wanted the reader to know that no one could seriously describe him as a Luddite—as if anyone had ever seriously thought to do so. But now, he was forced to admit, his general love of his technology had been dealt a cruel blow. Recently he had read a book called *The Age of Spiritual Machines* by his good friend, the MIT computer scientist Ray Kurzweil. A long passage in that book planted the seeds of doubt in Bill Joy. It read:

> First let us postulate that the computer scientists
> succeed in developing intelligent machines that

Eventually a stage may be reached at which the decisions necessary to keep the system running will be so complex that human beings will be incapable of making them intelligently. At that stage the machines will be in effective control. People won't be able to turn the machines off, because they will be so dependent on them that turning them off would amount to suicide.

On the other hand it is possible that human control over the machines may be retained. In that case the average man may have control over certain private machines of his own, such as his car or his personal computer, but control over large systems of machines will be in the hands of a tiny elite—just as it is today, with two differences. Due to improved techniques the elite will have greater control over the masses; and because human work will no longer be necessary the masses will be superfluous, a useless burden on the system. If the elite is ruthless they may simply decide to exterminate the mass of humanity. If they are humane they may use propaganda or other psychological or biological techniques to reduce the birth rate until the mass of humanity becomes extinct, leaving the world to the elite. Or, if the elite consists of softhearted liberals, they may decide to play the role of good shepherds to the rest of the human race. They will see to it that everyone's physical needs are satisfied, that all children are raised under psychologically hygienic conditions, that everyone has a wholesome hobby to keep him busy, and that anyone who may become dissatisfied undergoes "treatment" to cure his "problem." Of course, life will be so purposeless that people will have to be biologically or psychologically engineered either to remove their need for the power process or make them "sublimate" their drive for power into some harmless hobby. These engineered human beings may be happy in such a society, but they will most certainly not be free. They will have been reduced to the status of domestic animals.

Joy read along, with a growing sense of foreboding: he couldn't say he disagreed. Then he turned the page and discovered who had written it. Not Ray! It wasn't even some respected fellow member of the techno-elite. It was Theodore Kaczynski, a.k.a the Unabomber. "Like many of my colleagues," wrote Joy, "I felt that I could easily have been the Unabomber's next target . . . [but] as difficult as it is for me to acknowledge, I saw some merit in the reasoning of this single passage."

The threat Joy foresaw came indirectly from the Internet, or at any rate from the democratization of knowledge. He could now see a future when man would be able to create invisibly small robots—nanobots they were called. These robots could be designed to perform malicious deeds, such as eating the entire earth's crust in a few days. They'd also be able to reproduce without human assistance. The knowledge of how to create such self-replicating robots would be widely available to any psychotic, via the Internet. Or the evil deed might happen accidentally by some well-meaning techno-elite, in a lab experiment gone wrong. In either case the threat of tiny earth-eating robots set loose among us—known in the trade, apparently, as "the gray goo problem"—was at hand. As Joy wrote, "I think it is no exaggeration to say we are on the cusp of the further perfection of extreme evil, an evil whose possibility spreads well beyond that which weapons of mass destruction bequeathed to the nation-states, on to a surprising and terrible empowerment of extreme individuals." The main question on Joy's mind was the precise likelihood of a gray goo spill. "The philosopher John Leslie has studied this question," Joy wrote, "and concluded that the risk of human extinction is at least 30%, while Ray Kurzweil [him again] believes that we have 'a better than even chance' of making it through. . . ."

The article was an instant sensation. It was written up in the newspapers in reverential tones, and Joy himself went on serious-minded television shows to discuss it and soon signed a big book contract that would enable him to scare the hell out of a lot more people. Everyone at once seemed to understand: this was serious stuff. You might wonder how anyone could "calculate" the risk of human extinction. You might sense that that sentence about extreme evil could only be spoken in jest—say, by Austin Powers's nemesis, Dr. Evil. You might even sense that the minute some robot even looked at some poor guy sideways was the moment mankind reached for the screwdriver. But you were not to laugh. To laugh was not merely to increase the likelihood that one day you would find yourself on the gray goo lunch menu. To laugh at Bill Joy's premonition was to hold yourself up as one of those shortsighted souls destined to be plowed under by the many things your retinas failed to detect. If the thought of gray goo failed to get you thinking seriously, you needed to think again.

Or not. Joy's article wasn't really an invitation to rational thought. He wasn't seeking to give anyone permission to think about anything. His purpose was clear from his technique—that is, the manner in which he tried to establish the credibility of his sudden deep fear of the future. He did not bother to make himself an expert in gray goo. "I enter this arena as a generalist," he wrote, to acknowledge that he was flying by radar. Actually, that isn't all he said. What he wrote in his *Wired* article to deflect the obvious question—*What do you know about gray goo anyway?*—was, "As an architect of complex systems, I enter this arena as a generalist." I wondered how many readers knew what was meant by "an architect of complex systems." Two? Three? What he really meant to say was:

> As a member of the techno-elite, I should be
> above the critical scrutiny of the general reader.

At any rate, gray goo was not his field. He was just a software guy with some buddies who might or might not know the future of gray goo. Just as Danny Hillis never bothered to find out whether the yarn about the oak beams was actually true, Joy satisfied himself with knowing perhaps a bit more about gray goo than you and me. That is, he relied heavily (and safely, as it turned out) on the reader's belief that anyone smart enough to be called "the Edison of the

Internet" and to design something called "complex systems" must surely know what he was talking about, no matter what he was talking about.

Actually, the force of Joy's whole argument turned on people who might or might not be authorities on the subject at hand (it is unclear from the essay whether there even exists such a thing as an authority on gray goo), but who were, for what it was worth, and it was assumed to be worth a lot, the close friends of Bill Joy, whom he found to be, generally speaking, smart. In other words, he was doing what I have done for the last two hundred pages, only he was pretending to be doing a lot more. The author has a hunch, goes and talks to some people about his hunch. From these talks he concludes that he has an obligation to the masses to speak out. I kid you not. The man actually trundled around the Unabomber's manuscript to show to his friends. In his journey he discovered that some of these people shared his premonitions of doom. And so, he wrote, "I decided it was time to call my friend Danny Hillis. . . . I respect Danny's knowledge of the information and physical sciences more than that of any other single person I know. Danny is also a highly regarded futurist. . . . So I flew to Los Angeles for the express purpose of having dinner with Danny and his wife, Pati."

And off he went to ruin people's dinner with gray goo. In the end Hillis told Joy that at least part of what he foresaw—robots taking over from humans—would indeed occur, but so slowly that when the time came, it would not seem so bad. That is, they agreed about the future (whew!) but agreed to disagree about how they felt about it. "It seemed that he was at peace with this process and its attendant risks," Joy concluded, "while I was not."

It was all so weird it was hard to know where to begin. The premise here was obviously that trained men of computer science will think more logically than most about subjects outside their narrow fields—or that people would believe they did. And, it was true, if you didn't pause to reflect, you found yourself carried along by the sheer ponderousness of it all. "Gray goo would surely be a depressing ending to our human adventure on earth," Joy wrote. How true that is. On the other hand, what end to the human adventure on earth would not be depressing?

From the unspoken premise it followed that men of computer science were more qualified than most to think clearly about problems that are, in the end, political. After all, once you posit extreme evil you create a need for some kind of political response. One solution to the gray goo problem, Joy said, was for human beings to exit the planet Earth. Then again, he reflected, the gray goo will probably just follow us into space. That leaves but one final solution: stop the research that will lead inexorably to gray goo. But who will stop the research? Well, there's really only one answer to that question. Joy was calling, inevitably, for some central authority to drastically curtail the freedom of the individual to pursue knowledge. The idea of progress we now live by needed to be rethought. As unthinkable as that was, he wrote, "if open access to and unlimited development of knowledge henceforth puts us all in clear danger of extinction, then common sense demands that we reexamine even [our] basic, long held beliefs." That's his main point: We need to stop progress before progress stops us.

When I put down Joy's essay I thought: it was exactly how some old peer of the realm might have behaved had he found himself troubled by some new development. It was exactly what my father and his gentleman-lawyer friends might have done. Call the old boys in the network. Talk it over. Build consensus. Reach a conclusion that satisfies the old boys, and then call directly on political authority to take care of the problem. Assume everyone else will bow to the old boys' wisdom.

When highly self-conscious, highly intelligent, perfectly nice men chuck the principles on which they have built their careers and reinvent themselves as qualified enemies of their own idea of progress, it is as disconcerting in its way as gray goo on the kitchen floor. You see it and you know something is up. Hillis and Joy are trying to tell us something but they don't know how to say it.

The signature trait of their uncharacteristically irrational essays was how *personal* they were. Hillis planted his acorns in the style of Walt Whitman; Joy paused in his discussion of gray goo to tell us about his grandmother's moral guidance, and the effect of the early *Star Trek* episodes on his young mind. The writers wanted to be loved and admired not for their brief, ephemeral contributions to computer technology but for *themselves*. But that, they must know in their solar plexus, was increasingly a false hope. The success of those selves depended on their remaining in a childlike state of perpetual flux, waiting to jump the next time technology signaled a new direction. And suddenly they sense they don't quite have it in themselves to do this. They've grown up. And they know it. And they know what that means.

Go back to the passage from the Unabomber's Manifesto that transformed Bill Joy. Where it says "machine" plug in "youth," and where it says "humans" plug in "grown-ups."

Once you've done that the passage works much better as an explanation why a middle-aged technologist might rebel against his own system. The middle-aged technologist knows that somewhere out there some kid in his bedroom is dreaming up something that will make him obsolete. And when the dream comes true, he'll be dead wood. One of those people who need to be told to get out of the way. Part of the process.

You can't stop time. You can only simulate it, by stopping change. There is no solution to the lack of mercy toward the aged flushed out into the open by the Internet. The state of mind demanded by a world that quests after ever more rapid technical change is alien to anyone over forty. The best one can aim for is to be a case of arrested development and remain forever a child, fantasizing about clocks that tick for ten thousand years and gray goo that eats you in the night. Appealing as this willed arrest of one's development might sound, it is no celebration of childhood. It is a celebration only of the aspects of the childhood that have a market value: the child's gift for coping with the havoc he wreaks, and the child's ability to walk away from aspects of self without feeling like an amputee. The Internet allows a person to paint a picture of himself the world will value. But it has also helped to paint a picture of what the world values in a person. The effect is no less grotesque than one of those Spanish royal portraits in which the artist, needing to buttress the authority of some child monarch, renders his subject as a tiny adult.

In his hostility toward the future I found Bill Joy less compelling than Greg Lebed. Right up till the end of my trip I remained haunted by something Greg had said to me one night at his family's dinner table. As usual he'd spoken in anger. But there was always an honesty in Greg Lebed's rage; he knew what it felt like for the world to change in ways that were inimical to his own interests, and he wasn't going to pretend that he liked the feeling or that he could actually do something to forestall his own obsolescence. At any rate, in one of these hot moments, he'd pointed at his son and said, "As soon as that computer come into the house, that was it, childhood ended." After he said it, his wife pretended he had said nothing, his son rolled his eyes, and his daughter hid behind a mortified smile. The moment passed. But later I wrote to Jonathan and asked him if he thought there was any truth in what his father had said—that Jonathan had more or less quit his own childhood. Late one night Jonathan wrote back. Uncharacteristically, he decided not to fight the charges. Instead he simply explained his motive for committing the crime:

> The fact of the matter is that people spend most of their time as an adult. . . . While I don't think that it is good to miss your teenage years, and I don't think that I did miss much of my time as a teenager, I feel that it is very important to focus on the future right now.